Importing Agency

FOR FOREIGN PUBLICATIONS ON NATURAL HISTORY AND NATURAL SCIENCE.

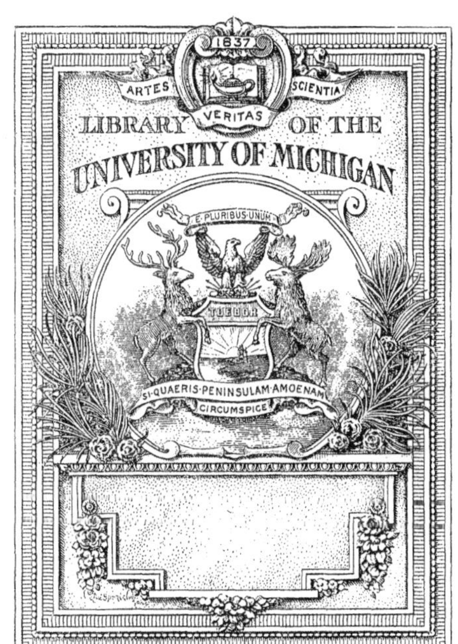

G ... ers from
the ... h Works
on N ... criptions
for ... y would
invi ... Mr. R.
HAI ... London,
any ... obtained.
Con ... ove sub-
jects ... re some
of t ... ouses:

TI ... HENRY
... resentative
Scien ... æontology,
Chem ... ics, Photo-
graph ... putting the general reader *au courant* with the progress of Science at home and abroad during the quarter which has elapsed previous to publication. The Journal furnishes an amount of scientific information, in a popular and yet exact form, which cannot be found in any other English Periodical.

Published Quarterly; price $5 a year; $1.25 a number. Complete sets and back numbers supplied.

II.

THE MONTHLY MICROSCOPICAL JOURNAL.

This Journal is devoted exclusively to the interests of Microscopical Science in the widest and most accurate sense of the term. It contains not only the proceedings of the Royal Microscopical Society, but also embraces communications from the leading Histologists of Great Britain, the Continent, and America, with a comprehensive *résumé* of the latest Foreign Inquiries, Critical Reviews and Short Notices of the more important works, Bibliographical Lists, and Descriptions of all new and improved forms of Microscopes and Microscopic Apparatus; Correspondence on all matters of Histological Controversy; and finally, a Department of "Notes and Queries," in which the student can put such questions as may elicit the special information he desires to obtain.

Published Monthly; price per year $9.00; per number 75 cents. Complete sets, comprising an exhaustive Record of the progress of Microscopy, copiously illustrated, 7 vols., 8vo, cloth, $36.

III.

THE FLORAL MAGAZINE. New Series, enlarged to Royal 4to. Figures and Descriptions of Select New Flowers for the Garden, Stove, or Conservatory. By the Rev. H. H. DOMBRAIN.

Monthly, with Four Colored Plates, $1.75; Annual Subscription, $21.

IV.

THE BOTANICAL MAGAZINE. Figures and Descriptions of New and Rare Plants of interest to the Botanical Student, and suitable for the Garden, Stove, or Greenhouse. By Dr. J. D. HOOKER, F.R.S.

Monthly, with Six Colored Plates, $1.75; Annual Subscription, $21.

FLORA.

THE NATURAL HISTORY OF PLANTS. By H. BAILLON, President of the Linnæan Society of Paris, Professor of Medical Natural History and Director of the Botanical Garden of the Faculty of Medicine of Paris. Translated by MARCUS M. HARTOG, B. Sc. Lond., Scholar of Trinity College, Cambridge. Vols. I. and II., with 800 Wood Engravings, $12.50 each.

ICONES PLANTARUM. Figures, with Brief Descriptive Characters and Remarks, of New and Rare Plants, selected from the Author's Herbarium. By Sir W. J. HOOKER, F.R.S. New Series, Vol. V. 100 Plates, $15.50.

SELECT ORCHIDACEOUS PLANTS. By ROBERT WARNER, F.R.H.S. With Notes on Culture, by B. S. WILLIAMS. Folio, cloth gilt. Second Series, Parts I. to X., each, with 3 Colored Plates, $5.25.

THE TOURIST'S FLORA;
A Descriptive Catalogue of the Flowering Plants and Ferns of the British Islands, France, Germany, Switzerland, Italy, and the Italian Islands. By JOSEPH WOODS, F.L.S. Octavo, $9.00.

BRITISH GRASSES;
An Introduction to the Study of the Grasses found in the British Isles. By M. PLUES. Crown 8vo, 16 Colored Plates, drawn expressly for the Work by W. FITCH, and 100 Wood Engravings, $5.25.

FERNS.

EVERY KNOWN FERN.
Synopsis Filicum; including Osmundaceæ, Schizæaceæ, Marattiaceæ, and Ophioglossaceæ, accompanied by Figures representing the essential characteristics of each Genus. By the late Sir W. J. HOOKER, K.H., and JOHN GILBERT BAKER, F.L.S., Assistant Curator of the Kew Gardens. Price $11 plain; colored by hand, $14.

FERNS, BRITISH AND FOREIGN.
Their History, Organography, Classification, Nomenclature, and Culture, with Directions showing which are the best adapted for the Hothouse, Greenhouse, Open Air Fernery, or Wardian Case. With an Index of Genera, Species, and Synonyms. By JOHN SMITH, A.L.S., late Curator of the Royal Gardens, Kew. With 250 Wood Cuts. Crown 8vo, cloth, fully illustrated, price $3.

Mr. Smith is acknowledged to be one of the first authorities on Ferns, having been engaged nearly half a century in arranging them at Kew.

GARDEN FERNS;
Colored Figures and Descriptions, with Analysis of the Fructification and Venation, of a Selection of Exotic Ferns, adapted for cultivation in the Garden, Hothouse, and Conservatory. By Sir W. J. HOOKER, F.R.S. Royal 8vo, 64 Colored Plates, $21.

INSECTS.

HARVESTING ANTS AND TRAP-DOOR SPIDERS.
Notes and Observations on their Habits and Dwellings. By J. T. MOGGRIDGE, F.L.S. Colored Plates, $5.25.

BRITISH INSECTS.
A Familiar Description of the Form, Structure, Habits, and Transformations of Insects. By E. F. STAVELEY, Author of "British Spiders." Crown 8vo, with 16 beautifully Colored Steel Plates and Numerous Wood Engravings, $7.

THE PREPARATION AND MOUNTING OF MICROSCOPIC OBJECTS.

THI

A

OF

MICROSCOPIC OBJECTS.

BY

THOMAS DAVIES.

Second Edition.—Greatly Enlarged.

EDITED BY JOHN MATTHEWS, M.D., F.R.M.S.,

Vice-Pres. Quekett Microscopical Club.

NEW YORK:

WM. WOOD & CO.

G. P. PUTNAM'S SONS.

1874.

PREFACE TO THE SECOND EDITION.

THE reception accorded to this work has been so favourable as to induce the Publisher to issue a second edition, in which such new matter should be embodied as the progress of Microscopic science might require. He therefore applied to the Author, but he found to his regret that the state of Mr. Davies's health was such as to forbid his undertaking the labour. He had, however, collected many valuable notes and memoranda, which he was willing to place at the disposal of any gentleman who might be selected to edit the work. The Publisher then consulted the present Editor, who, after some hesitation, consented not only to use his best efforts with the ample materials placed at his disposal, but also to make such additions as his experience might suggest in extension of the usefulness of the book to a new class of readers,—the Medical Student, and the Junior Medical Practitioner. To this end, besides other matter, a brief prefatory chapter has been added, embracing the elements of preliminary histological manipulation. While claiming the indulgence of the elders of his profession,—the Editor feels that the best and truest apology for this treatise, its *raison d'être*—in fact, may be found in the words of its concluding paragraph, to which the reader is now courteously referred.

4, Mylne Street, Myddelton Square, E.C.

October, 1873.

PREFACE TO THE FIRST EDITION.

IN bringing this Handbook before the public, the Author believes that he is supplying a want which has been long felt. Much information concerning the "Preparation and Mounting of Microscopic Objects" has been already published; but mostly as supplementary chapters only, in books written professedly upon the Microscope. From this it is evident that it was necessary to consult a number of works in order to obtain anything like a complete knowledge of the subject. These pages, however, will be found to comprise most of the approved methods of mounting, together with the results of the Author's experience, and that of many of his friends, in every department of microscopic manipulation; and as it is intended to assist the beginner as well as the advanced student, the very rudiments of the art have not been omitted.

As there is a diversity of opinion as to the best mode of proceeding in certain cases, numerous quotations have been made. Wherever this has been done, the Author believes that he has acknowledged the source from which he has taken the information; and he here tenders his sincere thanks to those friends who have so freely allowed him to make use of their works. Should, however, any one find his own process in these pages *unacknowledged*, the Author can only plead oversight, and his regret that such should have been the case.

WARRINGTON

THE

PREPARATION AND MOUNTING

OF

MICROSCOPIC OBJECTS.

CHAPTER I.

INTRODUCTION.

THIS work having been written chiefly to help students, the writer does not venture to affirm of it that it is by any means complete or exhaustive. The art of microscopic manipulation is progressive, and it is scarcely possible, therefore, to say of a work on the subject, that it holds all that is known at any given time. It is an art, too, which is so inextricably mixed up with the highest branches of scientific inquiry, that new modes of investigation are daily devised by the acutest intellects, and with these it is very difficult for a writer to keep pace.

It is a well-nigh hopeless task to attempt to teach such modes of inquiry by precept, yet it is felt that some short account of them may reasonably be expected here. Reference is now made more particularly to the practical part of human and comparative histology. As this is not a treatise on histology, but is devoted mainly to the methods of preserving the results of researches in that science, it is scarcely possible to indicate to the student how he shall proceed in any given case; yet there are certain tests, reagents, and staining matters employed, with the uses and effects of which he should be familiar, so as to be able to speak with some degree of certainty of the nature of the tissues demonstrated by them.

It is now, therefore, intended to give the reader a list of these aids, arranging them according to the effects which it is desired to produce. Stricker observes, "that it is to be borne in mind that it is impossible to say of any fluid that it constitutes an indifferent, *i.e.*, neutral, medium for fresh tissues of all kinds. In *all* instances we must be prepared for changes taking place." He gives, however, a list of fluids to which structures are generally most indifferent, *i.e.*, in which least alteration may be detected under examination while fresh, viz.:—

1st. Fluid of the aqueous humour.

2nd. The serum of the blood.

3rd. Amniotic fluid, very fresh, in which a little iodine has been dissolved, making it of a faint yellow tint.

4th. Very dilute solutions of neutral salts, such as phosphate and acetate of soda and potash, &c.

It is scarcely within the power of any one observer to have largely used or tested the whole of the processes hereinafter to be mentioned. The writer therefore freely admits his obligations to the treatises of Drs. Beale and Carpenter, Mr. Quekett and Mr. Fownes, as well as to those of Stricker, Frey, Klein, Schultze, Kühne, Deiters, Leber, and others, many of whose processes he has personally verified, and of whose manuals, especially those of Beale, Stricker, and Frey, the student is advised to possess himself. He believes also, from his own early experiences, that some short rationale of the intentions of the processes and means of investigation used by well-known workers may be acceptable to the student, in repeating their experiments before embarking in any of his own.

These materials and methods may be divided, then, and described according to their effects somewhat as follows; and it is in the judicious selection of each one or more of them that the tact and discretion of the student will best be shown. He should bear in mind, too, that the same structure may well be submitted to various modes of inquiry, and that possibly new modes may occur to him which, though they may not serve to prove anything directly, may yet become negative proofs.

1st. Such tests and agents as render transparent or translucent some tissues but not others adjacent, or make some more conspicuous than others without colouring them, or at least but faintly.

2nd. Staining materials or fluids, which colour either all the tissues to be displayed, or some particular part or parts of them, thus making such tissues or parts more conspicuous when subsequently examined or preserved in a colourless medium.

3rd. Hardening agents or solutions, by the effect of which tissues naturally so soft as to break down or be otherwise unmanageable under manipulation, are made firm enough for section, or for such examination as may suffice to to discriminate (or "differentiate") their parts, without either disturbing or confusing their structural relations.

4th. Softening agents of animal and vegetable tissues.

5th. Solvents of the same.

6th. Solvents of calcareous matter.

7th. Solvents of siliceous matter.

8th. Solvents of oily and fatty matters.

9th. Polarized light, by the agency of which structures and organs may often be optically differentiated as a preliminary to other modes of investigation.

10th. Electricity and heat.

11th. The moist chamber.

In dealing with structures by means of the agents comprised under the first division of our list, a very frequent and necessary preliminary is the teasing out or separation of fibres by means of two sharp needles set in convenient handles. But it must be remembered that an appearance of structure, where it does not really exist, may easily be thus produced. It is often necessary, also, that the object shall have been macerated in water, or some other agent, for so long a time as may be required to loosen or dissolve the connective tissue. It is of these agents that we shall presently have to speak in detail. greater or less according to their relative importance. Boiling or steaming may often be employed with advantage. It must not be too hastily concluded that, because there is nothing at first visible, there *is* therefore nothing to be seen. There are many important tissues which are apparently structureless, or homogeneous, which yet are possessed of such diverse elements as absolutely to require some process by which they may be optically or visually *differentiated*, if one may use such a phrase, *i.e.*, discriminated from the neighbouring tissues or organs. It is thus that their proper uses and purposes in relation to the whole organism may be correctly indicated or inferred, their histological nature decided, and their physiological relations and connections established beyond doubt.

The student is also very emphatically cautioned against the use of objectives of very wide angle, as well as of deep eye-pieces. In the former case, the relations of structures to each other can never be well made out, since it is impossible to get a focus of any depth (*i.e.*, of all the structures involved), in one view, because the objects in one plane only can be clearly seen, the rest, either above or below, being more or less out of focus, and therefore hazy and indistinct. This objection applies far less to those of lesser angle, which are therefore the best for histological purposes. In the latter case, we have nearly the same defect to contend with, viz., that surface markings only, or mostly, can be seen clearly (not to speak of the loss of light). The use of the draw-tube is the true remedy for this.

1st Division.

Under our *first head* may be ranged the following:—

Acetic acid.
Liquor potassæ and sodæ.
Concentrated sulphuric and hydrochloric acids (the latter saturated with chlorine).
Tannin.
Lime and baryta water.
Oxalic acid.
Nitric acid with chlorate of potash.
Ammonia.
Alcohol.
Iodine.
Glycerine (?)
Phosphoric acid (tribasic).

Acetic acid more or less dilute, *e.g.* one part to five of water, after a sufficiently prolonged immersion, renders *transparent* the following tissues, without in general destroying their connective tissue:—some *muscles* (of the frog (Kölliker).) cell-walls generally (not the nucleus), epithelial structures, *white* fibrous tissue. Dr. Beale says that yellow fibrous tissue is unaltered by it. Many kinds of formed material, sections of preparations which have been hardened by alcohol. Dr. Beale also says that it dissolves granular matter composed of albuminous material, and that many tissues are quite insoluble in it, though they are not rendered opaque by it. Acetic acid renders some tissues transparent by dissolving out the phosphate or carbonate of lime, which they may contain, but it has no similar solvent power over oxalate of lime. Parts which are unaffected by this acid are then made more conspicuous.

Liquor potassæ and liquor sodæ act in much the same way, according to the degree of their dilution, but on different structures. Albuminous tissues, epithelium, &c., are either dissolved by them or rendered so transparent as not to obstruct the view of the subjacent structures.

Concentrated sulphuric and hydrochloric acids, used co..., cause epidermic structures to swell up, so that their cells may be easily separated.

Tannin, dissolved in water or rectified spirit of wine, hardens gelatinous and albuminoid tissues: it also makes them shrink. Its solution in water has been used, as mentioned in another part of this treatise, as an injection preliminary to one of coloured gelatine, to prevent extravasation through the walls of the blood-vessels. It also colours the tissues a fawn-colour, or a very faint brown.

Lime water and baryta water, especially the latter, will, according to Rollet, dissolve the animal cement by which the fibres of connective tissue are held together. After a few days' soaking such tissue, as well as tendon, may easily be teased out by needles.

Oxalic acid, in a cold saturated solution (1 acid, 15 water), according to Schultze, "causes connective tissues to swell up and become transparent, while those formed of albuminous substances become hardened and isolated. Extremely delicate elements of the body, such as the rods of the retina, &c., are thus well preserved."

Strong nitric acid mixed with chlorate of potash destroys connective tissue in a short time, and is therefore a good medium for isolating muscular fibres (Kühne).

Sulphuric acid, highly diluted (1 to 1,000 parts of water), used warm, gelatinizes connective tissue, and is also useful for the isolation of muscular fibres.

Strong hydrochloric acid dissolves the intercellular substances of organs abounding in connective tissue.

Ammonia acts on animal matters much in the same way as potash and soda.

Alcohol coagulates albuminous tissues, and makes them opaque. It corrugates most transparent membranes, and thus renders them more visible.

Finally, it may be affirmed that there often exists a need of making objects which are too dark more transparent by means of a fluid which permeates them *unequally*, so that the tissues are thereby as it were "differentiated," yet not altered in any material degree. This may be effected by solutions of gum, sugar, glycerine, and creosote, if the tissues are moist. If dry, then turpentine, Canada balsam, benzine or benzole, and the essential oils of cloves, anise, and cassia, may be employed.

2nd Division.

Under the second division of our subject come staining fluids.

Many of these will be found mentioned in the body of this work. They comprise carmine solutions, both acid and alkaline; aniline colours, indigo, carmine, hæmatoxyline, &c., formulæ for the use of which are given. To these Frey adds, blue tingeing by molybdate of ammonia, and double staining by carmine and picric acid.

A neutral solution of the molybdate of ammonia of the strength of 5 per cent. gives a blue tint to nerve-tissue, lymphatic glands, and ciliated epithelial cells, after maceration for 24 hours in the light.

For double staining by carmine and picric acid he recommends a mixture containing—

1 part creosote,
10 parts acetic acid,
20 parts water.

Soak the tissues in this solution while boiling for about a minute, then dry for two days. Make thin sections of them, immerse for an hour in water faintly acidulated with acetic acid, and then wash in distilled water. Next place them in a very dilute watery solution of ammoniacal carmine, wash again in water, and place in a solution of picric acid in water, the strength of which will vary according to circumstances. The sections are then to be placed on a slide, superfluous acid allowed to drain off, and a mixture of 4 parts creosote to 1 of old resinous turpentine dropped on them. In about half an hour they will become transparent, and may be mounted in Canada balsam.

"A peculiar effect is thus obtained. Epithelial and glandular cells, muscles, and the walls of vessels show a yellowish colour, with reddened nuclei, while the connective tissue is not coloured by the picric acid, and only presents the carmine colour."

Another mode of effecting the above is by adding to a saturated and filtered solution of picric acid in water, a strong ammoniacal solution of carmine, drop by drop, until neutralization takes place. Sections may be soaked in this solution, more or less dilute, for a sufficient time, and treated as in the previous method.

The other staining agents are:—nitrate of silver, osmic acid, chloride of gold, chloride of gold and potassium, protochloride of palladium. These are to be made into weak solutions in distilled water, in which the tissues, in section or otherwise, are to be placed, and then exposed to light for a sufficient time.

Leber recommends a mode of staining by Prussian blue, as follows:—Immerse the specimen in a weak solution of a protosalt of iron for five minutes, more or less, according to size or the thinness of the section. Then wash and move it to and fro for a few minutes in a 1 per cent. solution of ferro-cyanide of potassium until it assumes an intense and uniform blue colour. Then wash in water, soak in alcohol, and mount as usual. The effect is that of partial tingeing; the colour penetrates very deeply, and the tissue may be subsequently stained with iodine, carmine, or fuchsine. This method has been used for the cornea of the frog.

Iodine 1 part, with 3 of iodide of potassium, dissolved in 500 of water, may be used for tingeing of a brown colour animal cells, as well as all amyloid substances, animal or vegetable, sulphuric acid being added.

3rd Division.

Under the third division of our list may be ranged the following agents :—alcohol, solutions of chromic acid, bichromate of potash, hyperosmic acid, chloride of palladium, bichloride of mercury (in Goadby's solution), and tannin, or the substance may be dried in thin layers or small pieces, either spontaneously or in vacuo, or by carefully regulated heat; in some cases it may be boiled, or it may be frozen.

Alcohol is, on the whole, the best and most convenient of the hardening agents. It acts by abstracting water and coagulating albumen, and its uses as a preservative fluid *per se* are well known. It enters also into many of the preservative fluids, and is especially convenient and useful when it is desired to mount specimens quickly out of watery fluids in Canada balsam, without drying them previously. After a longer or shorter soaking in it, according to their size or thinness, preparations may be at once placed in turpentine, and then easily and speedily put up for examination in balsam.

Dr. Beale recommends a mixture of alcohol and a solution of caustic soda for the preservation of delicate tissues. He observes, "that alcohol alone tends to coagulate albuminous textures and render them opaque, at the same time that it hardens them. The alkali, on the other hand, will render them soft and transparent, and would dissolve them if time were allowed. These two fluids, in conjunction, harden the texture, and at the same time make it clear and transparent."

Chromic acid in solution, 0·25—0·5 to 1 and 2 per cent. of distilled water is much used. On account of its deliquescence, it is most conveniently kept in a saturated solution, which may be diluted as desired; and very often the weaker this solution the better. When it has had the desired effect on the tissue, the preparation should be removed into diluted alcohol, on account of the readiness with which fungi and confervoid growths are formed in chromic acid solutions. There are some precautions needed for perfect success with this agent, for which the reader is referred to Frey's "Microscopic Technology."

Bichromate of potash, in solutions of similar strength to those of chromic acid, may be used in the same way, but is far slower in producing its effect, and therefore inferior in the opinion of many. Stricker, however, says, "that it has the great advantage that tissues saturated with it do not become friable, and that the time occupied by this agent, as well as by the preceding, may be much shortened by removing the preparation into alcohol for twenty-four hours."

It is always advisable to divide the substances to be hardened into portions as small as convenient, since the larger often putrefy in the centre, though they harden at the surface. It is quite certain that many of the more delicate structures, such as the rods of the cochlea of the ear (Pritchard), those of the eyes of insects, &c., are better prepared with this than by the preceding agent. One great element of success in these two processes is, that the volume of the solution should be very large in proportion to the size of the object; another, that the action should be commenced with a weak solution, and continued with a stronger.

It sometimes happens that objects may be hardened too much by these solutions, though there is less risk by the bichromate of potash. In such cases Frey recommends that they be soaked in glycerine for a few days, and even that it be added to the solutions at first. He, with Deiters, Arnold, Schultze, and Kühne, claims for these solutions an effect of the most important kind, viz., that of "preserving the finest textural relations, while exerting a somewhat macerating action on them, so that very delicate organizations, especially in nerve tissues, may be made visible which were previously hidden, or not visible in examination of the fresh tissue."

Hyperosmic acid and chloride of palladium are sometimes used for this purpose also. Their solutions may contain from one-fifth to one-tenth per cent. of distilled water.

Bichloride of mercury acts, in hardening tissues (like most of the preceding, probably), by combining and forming an insoluble compound with their albuminoid elements. It is not much employed for this purpose, but is principally of use in certain preservative solutions mentioned elsewhere in these pages.

Tannic acid forms insoluble compounds with a great variety of organic and especially animal substances, as solutions of starch and gelatine, solid muscular fibre and skin, &c., which then acquire the power of resisting putrefaction. It scarcely colours animal membrane. Dr. Beale says that its action upon red blood corpuscles is "very peculiar." The solution used is three grains to an ounce of water. Other uses of tannin (tannic acid) will be found elsewhere in this work, and the intelligent student will easily thence infer its action and properties.

Drying may be effected either in a current of warm dry air, or under a bell-glass over sulphuric acid, or over a layer of parched oatmeal; or a cheap form of water bath may be employed, such as will be found described in this work. Another very speedy method is to soak the specimen in strong alcohol for a sufficient time, remove it, and expose to a current of warm dry air.

Boiling.—Tissues may be hardened by boiling in a fluid consisting of 8 parts water, 1 part creosote, and 1 part vinegar, for two or three minutes. They may then be laid out to dry. After two or three days they acquire a firmness admirably adapted for section; but if they remain too long uncut they become of a consistence unfit for that purpose. On the whole, boiling is not to be recommended, though Stricker says that it has its occasional uses.

Freezing may be employed for otherwise unmanageable structures, such as brain, spinal cord, &c. (though there seems to be an objection of a theoretical kind to this use of it, viz., that it may injure or alter the cells), or other tissues which will not admit the use of chromic acid, or which it may be desired to view under other aspects.

The writer has little or no experience of this plan, he therefore quotes from Frey as follows :—

"The preparation is allowed to freeze (by contact, it is presumed, with a freezing mixture or solution) until it assumes a consistency which will permit fine sections to be made with a *cooled* razor. The object is more convenient to handle if it is allowed to freeze on a piece of cork. Nerves and muscles have been treated in this manner with good results. Glands (salivary), livers, spleens, the lungs, skin, and the bodies of *embryos* (see Beale's process for the same in this work), also ganglia, afford excellent appearances. Indifferent (or neutral) media, such as iodine serum, are to be used in examining such sections. Or the preparation may be held in paraffine wax (diluted or not with oil), or tallow, which have been melted, and the object suspended or plunged in them until they cool, and the cooling may be carried further, if needed, by freezing."

In reference to this subject, Mr. Kesteven informs the author that he has found the paraffine composition more useful for brain than spinal cord. The former can be cut into any angular shape, and be so held steady for slicing; but the cord, being round, becomes loosened in its setting of wax (or paraffine), and revolves with the pressure of the knife. For either brain or cord he prefers hand-cutting with a very sharp razor, after the manner of Lockhart Clarke (see Mr. Kesteven's paper in St. Bartholomew's Hospital Reports). If *many* sections are to be made from a brain, machine-cutting saves much time. The razor should have some spirit of wine dropped on it, so as to prevent the sections adhering. The cutting machines are generally graduated (by a screw and index) on the upward movement, so as to enable one to judge of the thickness of the section; but as the brain substance and paraffine are both yielding to a certain extent, the reading must be taken with allowance.

4TH DIVISION.

This includes glycerine, liquor potassæ and sodæ, heat (as regards some substances), maceration (carried to incipient putrescence), nitric and chlor-hydric acids, either pure or dilute (in the case of bones, nails, &c.).

The writer is in doubt whether glycerine ought or not to be included under this section or the first, its uses and effects being so various and interesting. Indeed, there is scarcely any agent to which histology is more indebted for its present status and progress, since there is now no doubt that elementary tissue can be more readily discriminated in this medium—perhaps, too, *by* it—than any other. It has also the valuable property of preserving the tissues, if it be not too much diluted, and even then it is generally effectual if camphor water be employed as the diluent. The strongest and best glycerine should always be employed. The first effect on tissues immersed in it is that they shrink, owing to the abstraction of their water; but Dr. Beale speaks in the highest terms of its uses and advantages, and declares that the tissues gradually regain their original volume if left in it for a sufficient time. They then soften, and even swell up. His practice is first to immerse the specimen in *weak* glycerine solution, and then gradually to increase the density of the fluid. He recommends, also, "in order that tissues may be uniformly permeated with a fluid within a very short time after the death of an animal, that the fluid should come *quickly* in contact with every part of the texture." This, he says, may be effected in two ways, by

A. Soaking very thin pieces in the fluid;
B. By injecting the fluid into the vessels of the animal.

He thinks that these properties more particularly appertain to glycerine than to any other medium, and affirms that "cerebral tissues, delicate nervous tissues like the retina or the nerve-textures of the internal ear, may be saturated with it, *and dissection then carried to a degree of minuteness impossible in any other medium.* All that is required is, that the strength of the fluid should be increased very gradually until the whole tissue is thoroughly penetrated by the strongest that can be obtained;" and "that thus *very hard textures* may be softened, so that by gradually increasing pressure and careful manipulation exceedingly thin layers can be obtained, without the relation of the anatomical elements to each other being much altered, or any of the tissues destroyed." He also takes occasion to observe, "that tissues immersed in water are destroyed by even moderate pressure; but that in a viscid medium (such as glycerine or syrup) the requisite pressure can be borne not only without injury or impairment of the discrimination of their parts, but with advantage to their detail." One very great advantage which results from the use of glycerine for the preparation of textures is, that however they may swell in it after prolonged immersion, a sufficient soaking in water will always restore them to their normal condition. Another is, that on account of its very high refractive power, it is peculiarly fitted for the preparation of structures to be investigated by polarized light, with the same advantage as in the preceding case, that they are still amenable to all other modes of inquiry.

The caustic alkalies—potash, soda, and ammonia, are solvents of all animal textures except chitine, and perhaps bone. As in nearly all cases a softening action, with little or no alteration of tissue, precedes the solvent action, these agents, and especially the first two, have their uses. Under their influence "a condition is induced very favourable to the imbibition of water, which afterwards penetrates very rapidly, so that cells swell up and burst." They may be used either with or without heat, and more or less dilute. There is one disadvantage attending their use, that objects can with difficulty be preserved after soaking in them.

Heat, applied either by the aid of hot water or steam, or the sand-bath, or a bath of fusible metals, or of melted lead, is a very efficient means of softening horny substances, whalebone, &c., and rendering them plastic. Very thin laminæ of these substances may also be procured by the employment of a well-sharpened scraper, such as that used by cabinet-makers. This plan applies more to longitudinal than to transverse sections; yet even the latter may be obtained by fixing the object while soft in a piece of hard wood, and scraping both together. Long continued slow boiling softens and eventually disintegrates nearly all animal and vegetable tissues. Muscular fibre and many other textures may thus be isolated, such as spiral vessels, &c., in vegetables.

Prolonged maceration in water, for the preparation of anatomical structures, generally bony, is a process too well known to need description here. The addition of very dilute nitric, hydrochloric, and acetic acids is much employed for the separation of muscular fibres, both striated and smooth. Two or three days are required, or even more.

Nails may be softened very quickly by hot concentrated sulphuric acid—or, still better, by liquor potassæ, strength about 25 to 27 per cent.—so as to show isolated and distended cells by solution of the intercellular substance.

Bones are softened, *i.e.* decalcified, by boiling or, still better, by slow maceration in weak solutions of nitric and hydrochloric acids, by the action of which the phosphate and carbonate of lime may be entirely removed. This process isolates the animal matter, *i.e.* the osseine—sometimes miscalled gelatine—with all its peculiar fibres and processes. But bones may be treated in another way, so as to show or isolate the bone corpuscles with their processes, by removal or destruction of the intercellular substance. Though this can scarcely be called softening them, yet it may be most fitly mentioned here. For this purpose, a Papin's digester is necessary. When the boiling of bones has been for a long time carried on by means of one of these machines, they seem to be dissolved; but on examination a coarse powder, consisting of the isolated corpuscles and their processes, is found at the bottom of the vessel, which will amply repay the trouble of examination.

Teeth may be treated in the same manner as bones, except for the examination of the enamel, which is best effected by sections and grinding. For that purpose developing teeth should be chosen, as in them the enamel prisms are most easily isolated.

5TH DIVISION.

As the solution of animal and vegetable tissues generally means the confusion or destruction of their histological elements, not much can or need be said of it here, except that it may be as well to indicate the special solvents and tests of the special components of all tissues, since it is upon a correct knowledge and appreciation of the degrees and differences of the action of these, that effective histological research must chiefly depend.

Albumen, when pure, is nearly insoluble in water, wholly so when coagulated by heat. In dilute caustic alkali it dissolves with facility. Solution of nitrate of potash, acetic and tri-basic phosphoric acids, and pepsine, dissolve the purest form of albumen procured from white of egg.

Fownes observes, "that it must be remembered that a considerable quantity of alkali and very minute quantities of the mineral acids, prevent coagulation by heat, and that the addition of acetic acid, indispensable to the test by mercury, produces the same effect."

Fibrine of blood is insoluble in both hot and cold water, but is partly dissolved by long-continued boiling. Fresh fibrine, wetted with concentrated acetic acid, forms after some hours a transparent jelly, which slowly dissolves in water. Very dilute caustic alkali dissolves fibrine completely. Phosphoric acid produces a similar effect. Fibrine of flesh, which is not identical with that of blood (Liebig), is soluble in cold water containing one-tenth of hydrochloric acid.

Casein is only soluble in water in the presence of free alkali in very small quantities. It is partly soluble also in very dilute acids.

Gelatin, chondrin, and osseine are the result of the boiling of animal membranes, skin, tendons, and bones, respectively at a high temperature for a sufficient time. They are insoluble in cold water, but easily dissolved by the use of heat. Alcohol, corrosive sublimate in excess, nitrate of mercury, and, most characteristically, tannin, precipitate gelatine—the latter when it is very largely diluted.

"Skin and tendons contain a substance which resists the action of boiling water for many hours. It is insoluble in cold concentrated acetic acid, but by long-continued boiling in it, is gradually dissolved, and more easily in hydrochloric acid." (Fownes.)

Horny substance—keratin, found in hair, nails, feathers, and epithelium, is obtained by finely dividing them, treating them with hot water, and afterwards by boiling alcohol and ether. The horny substance is then very soluble in caustic potash.

Of bones we have already spoken.

It has been mentioned elsewhere in this work, that all the internal organs of insects may easily be dissolved out by boiling in liquor potassæ, leaving their external chitinous structures, limbs, &c., unaffected. But this is a proceeding much to be deprecated, for various reasons which it is scarcely necessary to give here. It is far better to treat them in another way, by which these organs may be examined *in situ*, at least to a very great extent, as will presently be shown.

The parenchyma of leaves and many other vegetable structures may be decomposed by prolonged maceration in water, and then easily be washed away. Nitric acid, varyingly diluted, will produce the same effect more speedily, the objects not requiring the same amount of bleaching subsequently. But by far the best and most speedy method is, to place them in the liquid manure tank of the gardener for a sufficiently long maceration. The results of this plan are exquisitely beautiful.

6TH DIVISION.

The proper solvents of calcareous animal matters are nitric, hydrochloric, and sulphuric acids. The earth of bones consists of a combination of two tribasic phosphates of lime, both of which are entirely soluble in nitric and hydrochloric acids. Sulphuric acid abstracts a part of the lime of bones, leaving a superphosphate—a substance much used in agriculture as a manure. Fluoride of calcium, existing in small quantity in bones, but in larger in the enamel of teeth (and of the ganoid scales of fish?), is decomposed by sulphuric acid, which combines with the calcium, allowing the hydrofluoric acid to fly off in a gaseous state. Carbonate of lime dissolves in nitric and hydrochloric acids. The shells of mollusca, and testæ of echinodermata, consisting principally of carbonate of lime, are also soluble in the same acids, as well as those of nummulites foraminifera, &c., which have been infiltrated with siliceous matter. These present the most beautiful "casts," which are exactly of the shape of the Sarcode body and canal system, thus enabling their internal organs to be studied with much accuracy. Dr. Carpenter says that they are of "wonderful completeness."

7TH DIVISION.

Silica is nearly altogether insoluble in water, but dissolves freely in strong alkaline solutions. Its only acid solvent is hydrofluoric acid. Its combinations with a larger proportion of alkali are soluble in water, and from such solutions silica may be precipitated in a gelatinous or colloid form by acids, or separated by dialysis, in the form of colloid silica. This substance may be used for procuring certain modifications of crystals of salts for the polariscope, such as sulphate of magnesia, sulphate of copper, boracic acid, sulphate of zinc, &c. In its combination with a smaller proportion of alkali, forming glass, it is attacked by hydrofluoric acid and its vapour, and advantage may be taken of this property to engrave names, numbers, &c., neatly upon slides, for classification in the cabinet. The glass to be engraved must be coated with an etching ground of oily varnish or wax, and the necessary writing effected upon it by a point, which must pierce through the protective material. A shallow basin, made by bending up the edges of a piece of sheet-lead, is then prepared, a little powdered fluor spar placed in it, and enough sulphuric acid added to form a thin paste. The glass is then placed in any convenient way over the basin, the waxed side downwards. A gentle heat is next applied, whereby the vapour of hydrofluoric acid is disengaged. This acts upon the glass exposed by the point in a very few minutes, removing a portion of its surface. The wax must then be removed by turpentine. If the lines which result are then rubbed over with any coloured varnish, and the varnish gently wiped off by a soft piece of rag, a sufficient portion will most probably remain in the etched marks to render them easily visible and legible. Of course it will be as well to prepare many slides in this way at once. It is not necessary to coat the whole surface of the slides with the protective varnish, if the leaden basin be covered with a thin piece of wood or sheet-lead perforated with holes slightly larger than the surface to be etched, over which holes the slides must be inverted for a sufficient time. This latter hint applies more particularly to finished slides requiring to be labelled.

8th Division.

The proper solvents of the fixed oily and fatty matters are ether, benzole (or benzine), turpentine, and the essential oils generally. Castor oil is nearly the only one which is soluble in alcohol, the rest being only slightly so. They are all capable of saponification with caustic alkalis, and so become indirectly soluble in soft or distilled water, otherwise they are wholly insoluble in it.

The volatile or essential oils mix in all proportions with fatty oils, and are wholly soluble in ether and alcohol.

Camphor dissolves in a only very small proportion in water, but freely in alcohol, ether, and strong acetic acid.

9TH DIVISION.

It is by no means intended to speak here of the general properties and uses of polarized light. But in relation to its special powers in the "differentiation" of tissues, there is very much to be learned. To be fitted for examination by this method, objects must be made more or less transparent or translucent; and in effecting this it is advisable, perhaps necessary, to employ media of high refractive power. Even when so prepared, it may be further necessary in some cases to employ selenite or mica films, still more to enhance their colour. Not the least indication can be afforded to the observer as to what colours he should employ generally, yet it is a matter of frequent observation that what are called "neutral tints" are to be preferred, such as result from the judicious use of *compound* selenite stages adjusted properly for that special effect.

The media most suitable for the preparation of objects to be examined in this way are glycerine, syrup, turpentine, dammar and benzole—or the latter alone, Canada balsam, and the essential oils. Of course sections must be made of tissues otherwise too thick. Of the advantages of employing the first of these we have already spoken; but to these must be added this important one, that it does not spoil the object for examination by other methods, if the glycerine be soaked out by maceration in water; and this is true also of syrup, though it is far less useful. For preparation by the other methods, tissues must have been soaked in alcohol, and then removed into the turpentine, &c.

For the examination of insects by polarized light, two preliminaries are necessary. Firstly, that they be made transparent or translucent by prolonged soaking in one of the above-named media, preferably in turpentine or the essential oils, or benzole. Secondly, that as in (most?) many of them their chitinous case is too deeply coloured for any amount of soaking to render them sufficiently transparent, some bleaching process should be premised. A formula for such a process may be found in another part of this work, where the preparation of the antennæ of insects is described. If that should not prove successful, some modification will easily occur to the student. Of course it is not all insects that can be treated in this way, the size and deep colour of very many quite preventing a good result; but when they have been successfully prepared by any of the methods of which we have spoken, it is then possible to discriminate their internal organs by the differences of colour which they present. The use of the binocular microscope, and of objectives of low angular aperture, will also much facilitate this mode of examination, by increasing the depth of focus, and enabling the organs to be seen more or less in connection with each other, even if they be superposed. It is also possible to examine the muscles of the limbs and bodies of insects, so as to decide upon their formation, origin, and insertion, and probable mode of action; and this is only one of many such uses. What a mistake must it be, then, to prepare insects for mounting by boiling in liquor potassæ, and so dissolving out their viscera, and squeezing them flat!

In the case of living insects, especially those of the more transparent salt and fresh water species, the results of their examination by polarized light are exquisitely beautiful and interesting, because their organs and circulation may be more clearly discriminated while in motion.

10th Division.

Electricity has been employed in histology partly for its electrolytic effects, but chiefly as a means of producing certain variations of temperature in objects under examination. Stricker says "that the tissues become altered by it as they would be were they subjected to the action of weak acids or alkalias," and he describes a rather complicated apparatus for this purpose, of which it is impossible to give an account here; but the author believes that most, if not all, of the same effects may be produced by the employment either of a thick plate of metal placed upon the stage, or of a thin water-bath, which may be heated by a spirit or gas flame, after the glass slide shall have been placed on it. They should both be properly fitted with thermometers.

Of the decomposition of salts by electricity, and their reduction to the metallic state, it is not necessary to speak here, but such effects are very beautiful, and the resulting crystalization may easily be watched.

Dr. Beale speaks very favourably of the inverted microscope devised by Dr. Lawrence Smith, U.S.A., by which objects may be viewed from their under instead of from their upper surface, and at the same time heated (or re-agents applied to them) without any risk of dimming or injuring the object-glass by vapours thus raised. The optical part is so fitted to the base that it may be drawn away from beneath the stage (to make room for the application of the lamp, or) for the sake of changing the powers.

11th Division.

It is evident that in all these plans an amount of evaporation is constantly going on, which will eventually dry and so spoil the object, unless obviated. Frey, therefore, describes a "moist chamber" invented by Recklinghausen for this purpose. It consists of a glass ring, more or less high, which has been cemented by its edge to a broad glass slide. A tube of thin rubber is then firmly fastened about the ring. The upper end of this tube is also fastened around the tube of the microscope. In order to keep the place thus enclosed saturated with moisture, some small pads of wetted bibulous paper, or pieces of elder pith also saturated with fluid, are to be enclosed with the object, which in this case need not be covered with thin glass in the usual manner. It is conceivable also that this apparatus may easily be converted into a gas chamber, by fixing two small, light vulcanized tubes into that which embraces the glass ring and the end of the microscope tube—one for the entrance, the other for the exit of the gas. This is a simpler and less costly plan than that devised by Stricker. Frey observes that it is most advantageous to use *immersion lenses* and the moist chamber with the hot stage.

CHAPTER II.

APPARATUS.

Before entering into the subject of the setting of Objects for the Microscope, the student must be convinced of the necessity of cleanliness in everything relating to the use of that instrument. In no branch is this more apparent than in the *preparation* of objects; because a slide which would be considered perfectly clean when viewed in the ordinary way is seen to be far otherwise when magnified some hundreds of diameters; those constant enemies, the floating particles of dust, are everywhere present, and it is only by unpleasant experience that we fully learn what *cleanliness is.*

An object which is to be viewed under the microscope must, of course, be supported in some way—this is now usually done by placing it upon a glass slide, which on account of its transparency has a great advantage over other substances. These "slides" are almost always made of one size, viz., three inches long by one broad, generally having the edges ground so as to remove all danger of scratching or cutting any object with which they may come in contact. The glass must be very good, else the surface will always present the appearance of uncleanliness and dust. This dusty look is very common amongst the cheaper kinds of slides, because they are usually made of "sheet" glass; but is seldom found in those of the quality known amongst dealers by the name of "patent plate." This latter is more expensive at first, but in the end there is little difference in the cost, as so many of the cheaper slides cannot be used for delicate work if the mounted object is to be seen in perfection. These slides vary considerably in thickness; care should, therefore, be taken to sort them, so that the more delicate objects with which the higher powers are to be used may be mounted upon the *thinnest*, as the light employed in the illumination is then less interfered with. To aid the microscopist in this work, a metal circle may be procured, having a number of different sized openings on the outer edge, by which glass slides can be measured. These openings are numbered, and the slides may be separated according to these numbers; so that when mounting any object there will be no need of a long search for that glass which is best suited to it.

When fresh from the dealer's hands, these slides are generally covered with dust, &c., which may be removed by well washing in clean rain-water; but if the impurity is obstinate, a little washing soda may be added, care being taken, however, that every trace of this is removed by subsequent waters, otherwise, crystals will afterwards form upon the surface. Sometimes, however, a certain greasiness is very obstinate upon the glass. It is then necessary to use a little liquor potassæ with a small piece of linen, rubbing the slide with some pressure, and then washing as before to remove all remains. A clean linen cloth should be used to dry the slides, after which they may be laid by for use. Immediately, however, before being used for the reception of objects by any of the following processes, all dust must be removed by rubbing the surface with clean wash-leather or a piece of cambric, and, *if needful*, breathing upon it, and then using the leather or cambric until perfectly dry. Any small particles left upon the surface may generally be removed by blowing gently upon it, taking care to allow no damp to remain. A very efficient remedy, also, is a mixture of equal parts of sulphuric ether and alcohol, with which the glass must be rubbed by the aid of a tuft of clean cotton-wool until no stain appears after breathing upon it. A strong infusion of nutgalls may be used in the same way, and is preferred by many to all other applications; or, a mixture of equal parts of alcohol, benzole, and liquor sodæ may be employed, which thoroughly and speedily cleanses glass from all traces of grease or balsam.

We have before said that any object to be viewed in the microscope must have its support; but if this object is to be preserved, care must be taken that it is defended from dust and other impurities. For this purpose it is necessary to use some transparent cover, the most usual at one time being a plate of mica, on account of its thinness; this substance is now, however, never used, thin glass being substituted, which answers admirably. Sometimes it is required to "*take up*" as little space as possible, owing to the shortness of focus of the object-glasses. It can be procured of any thickness, from one-fiftieth to one-two-hundred-and-fiftieth of an inch. On account of its want of strength, and probable defect of due annealing, it is difficult to cut, as it is very liable to "*fly*" from the point of the diamond. To overcome this tendency as much as possible, it must be laid upon a thicker piece, previously made wet with water, which causes the thin glass to adhere more firmly, and consequently to bear the pressure required in cutting the covers. The process of cutting being so difficult, especially with the thinner kinds, little or nothing is gained by cutting those which can be got from the dealers, as the loss and breakage is necessarily greater in the hands of an amateur. It is convenient, however, to have on hand a few larger pieces, from which unusual sizes may be cut when required.

If the pieces required are *rectangular*, no other apparatus will be required save a diamond and a flat rule; but if *circles* are wanted, a machine for that purpose should be used (of which no description is necessary here). There are, however, other contrivances which answer tolerably well. One method is, to cut out from a thick piece of cardboard a circle rather larger than the size wanted. Dr. Carpenter recommends metal rings with a piece of wire soldered on either side; and this, perhaps, is the best, as cardboard is apt to become rough at the edge when much used. A friend of mine uses thin brass plates with circles of various sizes "turned" through them, and a small raised handle placed at one end. The diamond must be passed round the inner edge, and so managed as to meet again in the same line, in order that the circle may be true, after which it may be readily disengaged. The sizes usually kept in stock by the dealers are one-half, five-eighths, and three-quarters inch diameter; but other sizes may be had to order.

For the information of the beginner it may be mentioned here that the price of the circles is a little more than that of the squares; but this is modified in some degree by the circles being rather lighter. If appearance, however, is cared for at all, the circles look much neater upon the slides when not covered with the ornamental papers; but if these last are used (as will shortly be described) the squares are equally serviceable.

As before mentioned, the thin glass is made of various thicknesses, and the beginner will wish to know which to use. For objects requiring no higher power than the one-inch object-glass, the thicker kinds serve well enough; for the half-inch the medium thickness will be required; while, for higher powers, the thinnest covers must be used. The "test-objects" for the highest powers require to be brought so near to the object-glass that they admit of the very thinnest covering only, and are usually mounted betwixt glasses which a beginner would not be able to use without frequent breakage; but if these objects were mounted with the common covers, they would be really worthless with the powers which they require to show them satisfactorily.

It may be desirable to know how such small differences as those betwixt the various thin glass covers can be measured. For this purpose there are two or three sorts of apparatus, all, however, depending upon the same principle. The description of one, therefore, will be sufficient. Upon a small stand is a short metal *lever* (as it may be termed) which returns by a spring to one certain position, where it is in contact with a fixed piece of metal. At the other end this lever is connected with a "finger," which moves round a dial like that of a watch, whereupon are figures at fixed distances. When the lever is separated from the metal which is stationary, the other end being connected with the "finger," of the dial, that "finger" is moved in proportion to the distance of the separation. The thin glass is, therefore, thrust betwixt the end of the lever and fixed metal, and each piece is measured by the figures on the dial in stated and accurate degrees. This kind of apparatus, however, is expensive, and when not at our command, thin glass may be placed edgewise in the stage forceps, and measured very accurately with the micrometer, or by the calliper eye-piece described by Dr. Matthews in No. 8, for October, 1869, of the Journal of the Quekett Microscopical Club.

Cleanliness with thin glass is, perhaps, more necessary than with the sides, especially when covering objects which are to be used with a high power; but it is far more difficult to attain, on account of the liability to breakage. The usual method of cleaning these covers is as follows:—Two discs of wood, about two inches in diameter, are procured, one side of each being perfectly flat and covered with clean wash-leather. To the other side of these a small knob is firmly fixed as a handle, or where practicable, the whole may be made out of a solid piece. In cleaning thin glass, it should be placed betwixt the covered sides of the discs, and may then be safely rubbed with a sufficient pressure, and so cleaned on both sides by the leather. If, however, the glass be greasy, as is sometimes the case, it must be first washed with a strong solution of potash, infusion of nutgalls, or any of the commonly used grease-removing liquids; and with *some* impurities water, with the addition of a few drops of strong acid, will be found very useful, but this last is not often required.

This method of cleaning thin glass should always be used by beginners; but after some experience the hand becomes so sensitive that the above apparatus is often dispensed with, and the glasses, however thin, may be safely cleaned betwixt the fingers and thumb with a cambric handkerchief, having first slightly damped the ends of the fingers employed to obtain firm hold. When the dirt is very obstinate, breathing upon the glass greatly facilitates its removal, and the sense of touch becomes so delicate that the breakage is inconsiderable; but this method cannot be recommended to novices, as nothing but time spent in delicate manipulation can give the sensitiveness required.

It has been before mentioned that ordinary glass sides are sometimes worthless, *especially for fine objects*, from having a rough surface, which presents a dusty appearance under the microscope. This imperfection exists in some *thin glass* also, and is irremediable; so that it is useless to attempt to cleanse it; nevertheless, care should be taken not to mistake dirt for this roughness, lest good glass be laid aside for a fault which does not really belong to it.

When any object which it is desired to mount is of considerable thickness, or will not bear pressure, it is evident that a wall must be raised around it to support the thin glass—this is usually termed a "cell." There are various descriptions of these, according to the class of objects they are required to protect; and here may be given a description of those which are most generally used in mounting "dry" objects, leaving those required for the preservation of liquids until we come to the consideration of that mode of mounting. Many have used the following slides. Two pieces of hard wood of the usual size (3 in. by 1 in.), not exceeding one-sixteenth of an inch in thickness, are taken, and a hole is then drilled in the middle of one of these of the size required. The two pieces are then united by glue or other cement, and left under pressure until thoroughly dry, when the cell is fit for use. Others substitute cardboard for the lower piece of wood, which is less tedious, and is strong enough for every purpose. This class of "cell" is, of course, fitted for opaque objects only where no light is required from below; and as almost all such are better seen when on a dark background, it is usual to fix a small piece of black paper at the bottom of the cell upon which to place them. For very small objects the grain which all such paper has when magnified detracts a little from the merit of this background; and lately I have used a small piece of thin glass covered on the back with black varnish, and placed the object upon the smooth untouched side; but a solution of the best Egyptian asphalt in benzole of moderate thickness may be painted on with this further advantage, that in mounting such opaque objects as foraminifera, &c., it will be sufficient to arrange them in the positions they are to occupy, when by slightly warming the slide they will adhere to the asphalt.

Another method of making these cells is as follows:—Two punches, similar to those used for cutting gun-wads, are procured, of such sizes that with the smaller may be cut out the centre of the larger, leaving a ring whose side is not less than an one-eighth of an inch wide. These rings may be readily made, the only difficulty being to keep the sides parallel; but a little care will make this easy enough. For this purpose close-grained cardboard may be conveniently used. It must have a well-glazed surface, else the varnish or cement used in affixing the thin glass cover sinks into the substance, and the adherence is very imperfect. When this takes place it is easily remedied by brushing over the surface of the cardboard a strong solution of gum or isinglass; and this application, perhaps, closes also the pores of the card, and so serves a double purpose. But, of course, the gum must be *perfectly* dried before the ring is used.

For cardboard, gutta-percha has been substituted, but cannot be recommended, as it always become brittle after a certain time, never adheres to the glass with the required firmness, and its shape is altered when worked with even a little heat. Leather is often used, and is very convenient; it should be chosen, however, of a close texture, and free from oil, grease, and all those substances which are laid upon it by the dressers.

Rings of cardboard, &c., have been rejected by persons of

great experience, because they are of such a nature that dampness can penetrate them. This fault can be almost, if not totally, removed by immersing them in some strong varnish, such as the asphalt varnish hereinafter mentioned; but they must be left long enough when affixed to the glass slide to become *perfectly dry*, and this will require a much longer time than at first would be supposed.

There has, however, been lately brought out what is termed the *ivory cell.* This is a ring of ivory-like substance, which may be easily and firmly fixed to the glass slide by any of the commonly used cements, and so forms a beautiful cell for any dry objects. They are made of different sizes, and are not expensive. Flat rings of brass turned down to the sizes of the circular discs of covering glass and of varying thickness are very neat and useful for mounting opaque objects: they can also be obtained in tin and zinc.

Some of our best microscopic men have stated that they have been frequently disappointed by an accumulation of encrusted matter upon the inner surface of thin glass used to cover the cell enclosing any dry object, and therefore use a shallow pillbox, made expressly for this purpose, which must be strongly cemented to the slide. For examination the lid must be removed, whilst it must be closed to protect the object from dust when laid aside. Another worker of experience recommends a cell in a mahogany slide, over which, by aid of a stud as on a pivot, a bone disk can be turned: this is termed, "Piper's Revolving Cover Slide," and can be procured at the opticians'.

Sometimes slides are used which are made by taking a thin slip of wood of the usual size (3 in. by 1 in.), in the centre of which is cut a circular hole large enough to receive the object. A piece of thin glass is fixed underneath the slide forming a cell for the object, which may then be covered and finished like an ordinary slide. This has the advantage of serving for transparent objects for which the before-mentioned wooden slides are unsuitable. A slight modification of this plan is often used where the thickness of the objects is inconsiderable, especially with some of the Diatomaceæ, often termed "test-objects." The wooden slide is cut with the central opening as above, and two pieces of thin glass are laid upon it, betwixt which the diatoms or other objects are placed, and kept in their proper position by a paper cover. This arrangement is a good one, insomuch as the very small portion of glass through which the light passes on its way to the microscope from the reflector causes the refraction or interference to be reduced to the lowest point.

A novice would naturally think the appearance of some of the slides above mentioned very slovenly and unfinished; but they are often covered with ornamental papers, which may be procured at almost every optician's, at a cost little more than nominal, and of innumerable patterns and colours. How to use these will be described in another place.

It is very probable that a beginner would ask his friend what kind of slides he would advise him to use. Almost all those made of wood are liable to warp more or less, even when the two pieces are separate or of different kinds; those of cardboard and wood are generally free from this fault, yet the slides, being opaque, prevent the employment of the Lieberkuhn. To some extent glass sides, when covered with ornamental papers, are liable to the same objection, as the light is partly hindered. And sometimes dampness from the paste, or other substance used to affix the papers, penetrates to the object, and so spoils it, though this may be rendered less frequent by first attaching the *thin* glass to the slide by some harder cement. Much time, however, is taken up by the labour of covering the slides, which is a matter of consideration with some. Certainly the cost of the glass slides was formerly great; but now they are reasonable enough in this respect, so that this objection is removed. It is, therefore, well to use glass slides, except where the thin glasses are employed for tests, &c., as above. When the thin glass circles are placed upon the slides, and the edge is varnished with black or coloured rings, the appearance of finish is perfect. The trouble is much less than with most of the other methods, and the illumination of the object very slightly impaired.

To varnish the edges of these covers, make circles of any liquid upon the glass slide, and perform any other circular work mentioned hereafter, the little instrument known as "Shadbolt's turntable" is almost indispensable. It is made as follows:—At one end of a small piece of hard wood is fixed an iron pivot about one-eighth inch thick, projecting half an inch from the wood, which serves as a centre upon which a round brass table three inches in diameter revolves. On the surface of this are two springs, about one and a half-inch apart, under which the slide is forced and so kept in position, whilst the central part is left open to be worked upon. The centre is marked, and two circles half an inch and one inch in diameter are usually deeply engraved upon the table to serve as guides in placing the slide, that the ring may be drawn in the right position. When the slide is placed upon the table underneath the springs, a camel-hair pencil is filled with the varnish, or other medium used, and applied to the surface of the glass; the table is then made to revolve, and a circle is consequently produced, the diameter of which it is easy to regulate. Mr. Hislop places two equidistant pins at opposite sides of the centre of the revolving plate, against which the opposing edges of the slip are made to bear, so that the instrument is self-centering. The springs are turned in contrary directions and are screwed on the pins, or the screws are made into the pins against which the sides of the slide bear.

The form of this "turntable" has been modified by many manipulators to suit their several wants. Almost all slides used are of nearly the same size—3 in. by 1 in.; and therefore the centres of all are equidistant from the edge. On this account one of my friends has a thin brass bar screwed upon the side of his turntable in such a position that the centres of the slides and table always coincide. The rings of varnish upon the slides and thin glass upon the cell are thus kept uniform. Dr. Matthews, a gentleman of no little experience, has given us an improvement as follows:—Take two "jaws" of the average thickness of a glass slide, $\frac{3}{8}$ inch wide, $2\frac{1}{2}$ long. Each of these is pivoted on the face of the turntable by a screw through its centre, each screw being placed exactly equidistant from the centre of the turntable, so that the jaws are separated by a space as wide as an average slide; *i.e.* a full inch. Outside of that space, on one side of the centre of one of the jaws, is a wedge fixed by a screw in such a way as to be capable of motion in the direction of its length by a slotted hole. This is all the machinery. AB and CD are the two jaws, E is the wedge. On placing a slip between the jaws they probably at first do not touch it. If the wedge be then pushed so as to approximate B to C, the jaws move on their centres, so that, however far B may be pushed towards (and moving) C, the other end of C—*i.e.* D—is moved *exactly* as much in the opposite direction until they approach near enough to grasp the slide by its edges. The length of the wedge must, of course, be such as to provide for about $\frac{1}{8}$ inch variation in the width of slides. It will readily be seen that the slip may be pushed in either direction excentrically lengthwise, so as to allow of the formation of any number of cells, all of which must needs be central as regards their width, if the instrument has been accurately made, which is a very easy matter. I have added also a rest for the hand, F, which may be turned aside on a centre at will, and which I have found to be a great convenience.

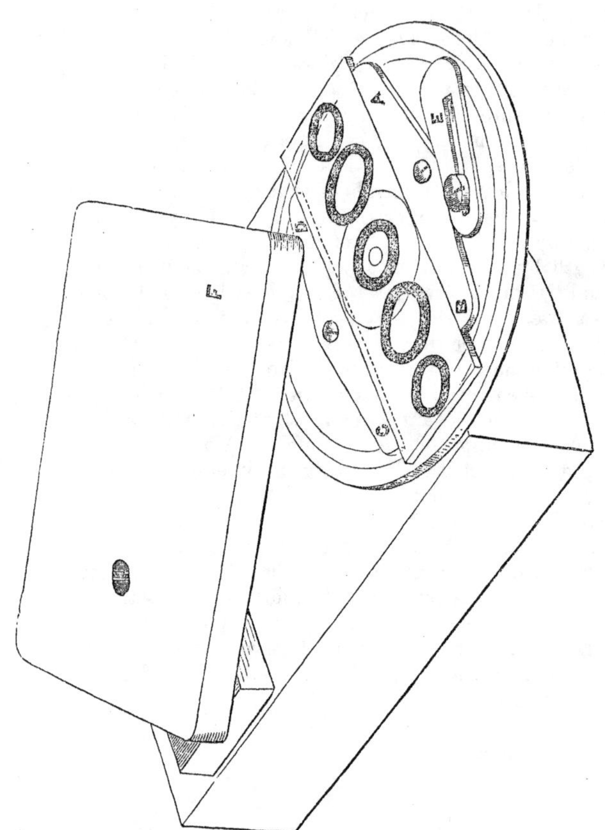

DR. MATTHEWS' TURNTABLE.

Mr. Spencer slightly modifies the above, using wood jaws and wedge, which the following engraving will best explain.

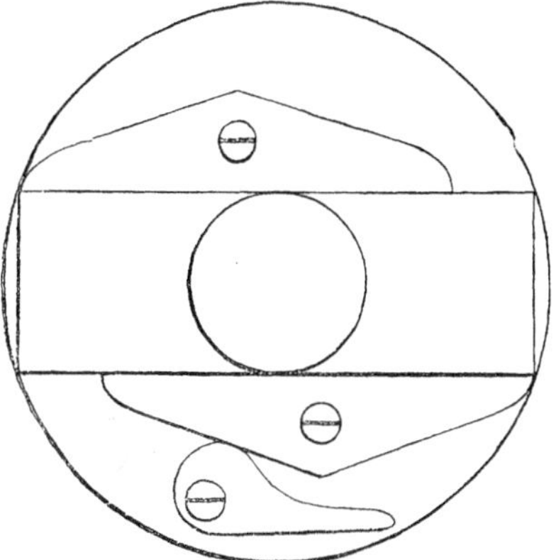

DR. MATTHEWS' TURNTABLE, TWO-THIRD SIZE.

Many objects for the microscope may be seriously injured by allowing the fingers to touch them—many more are so minute that they cannot be removed in this way at all, and often it is necessary to take from a mass of small grains, as in sand, some particular particle. To accomplish this, there are two or three contrivances recommended: one by means of split bristles, many of which will readily be found in any shaving-brush when it has been well used. The bristles when pressed upon any hard surface, open, and when the pressure is removed close again with a spring; but the use of these is limited. Camel-hair pencils are of great service for this, and many other purposes, to the microscopist. In *very* fine work they are sometimes required so small that all the hairs, with the exception of one or two finer pointed ones, are removed. A few of various sizes should always be kept on hand.

Equally necessary are fine-pointed needles. They are very readily put up for use by thrusting the eye end into a common penholder, so as to be firm. The points may be readily renewed, when injured, on a common whetstone; and when out of use they may be protected by being thrust into a piece of cork.

In laying out animal tissues that have been stained by nitrate of silver or chloride of gold, it is advisable to employ a small rod of glass drawn out to a point, as the use of a metallic point causes a deposit of gold or silver at the place of contact, which disfigures the preparation.

Knives of various kinds are required in some branches of microscopic work; but these will be described where dissection, &c., is treated at some length, as also various forms of scissors. In the most simple objects, however, scissors of the usual kind are necessary. Two or three sizes should always be kept at hand, sharp and in good order.

A set of glass tubes, kept in a case of some sort to prevent breakage, should form part of our fittings, and be always cleaned immediately after use. These are generally from six to ten inches long, and from one-eighth to a quarter of an inch in diameter. One of these should be straight and equal in width at both ends; one should be drawn out gradually to a fine point; another should be pointed as the last, but slightly curved at the compressed end, in order to reach points otherwise unattainable. It is well to have these tubes of various widths at the points, as in some waters the finer would be inevitably stopped. For other purposes the fine ones are very useful, especially in the transfer of preservative liquids which will come under notice in another chapter.

Forceps are required in almost all microscopic manipulations, and consequently are scarcely ever omitted from the microscopic box, even the most meagrely furnished; but of these there are various modifications, which for certain purposes are more convenient than the usual form. The ordinary metal ones are employed for taking up small objects, thin glass, &c.; but when slides are to be held over a lamp, or in any position where the fingers cannot conveniently be used, a different instrument must be found. Of these there are many kinds; but Mr. Page's wooden forceps serve the purpose very well. Two pieces of elastic wood are strongly bound together at one end, so that they may be easily opened at the other, closing again by their own elasticity. Through the first of these pieces is loosely passed a brass stud, resembling a small screw, and fastened in the second, and through the second a similar stud is taken and fixed in the first—so that on pressure of the studs the two strips of wood are opened to admit a slide or other object required to be held in position. The wood strips are generally used three or four inches long, one inch wide, and about one-eighth inch thick.

Again, some objects when placed upon the glass slide are of such an elastic nature that no cement will secure the thin glass covering until it becomes hard. This difficulty may be overcome by various methods. The following are as good and simple as any. Take two pieces of wood about two inches long, three-quarters wide, and one-quarter thick; and a small rounded piece one inch long, and one-quarter in diameter; place this latter betwixt the two larger pieces. Over one end of the two combined pass an india-rubber band. This will give a continual pressure, and may be opened by bringing the two pieces together at the other end; the pressure may be readily made uniform by paring the points at the inner sides, and may be regulated by the strength of the india-rubber band. These bands may be made cheaply, and of any power, by procuring a piece of india-rubber tubing of the width required, and cutting off certain breadths. Another very simple method of getting this pressure is mentioned in the "Micrographic Dictionary." Two pieces of whalebone of the length required are tied together firmly at each end. It is evident that any object placed betwixt them will be subject to continual pressure. The power of this may be regulated by the thickness and length of the whalebone. This simple contrivance is very useful.

Almost every scientific man, however, has his own model, and it may be as well to examine one or two of them. Mr. Goode uses the following: A, a piece of wood 8 in. long and ¾ in. thick. B, a spring made with thin iron wire. The end of the spring is driven into the table, as at C. A piece of ¼-in. iron wire is then run through the springs, which forms an axis to work upon, and also keeps them in their places. He inserts a pin at the side of the spring, so that it will fall on a given spot, and not rub the cover from side to side. The springs are made by binding the thin wire round the ¼-in. rod about four or five times.

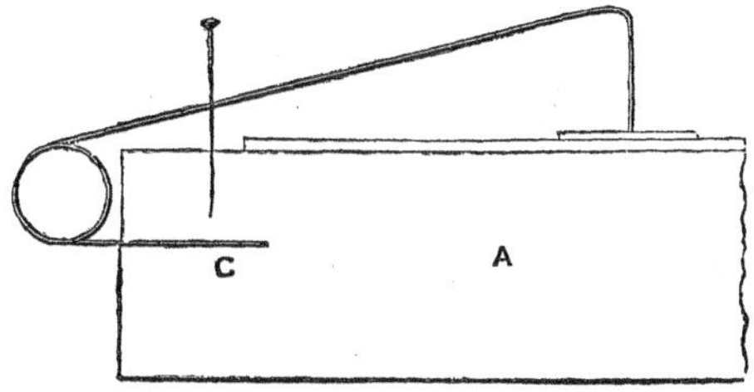

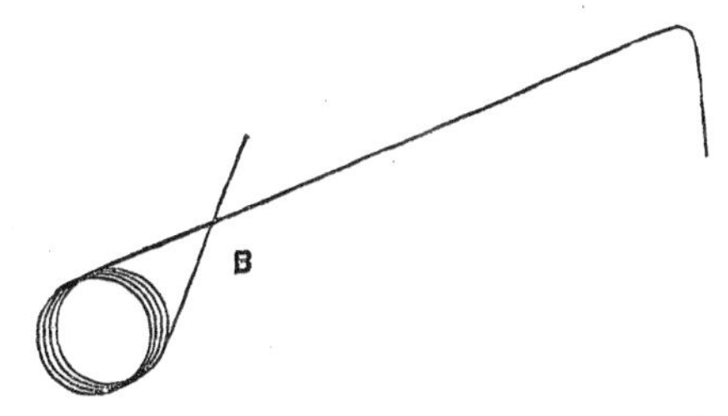

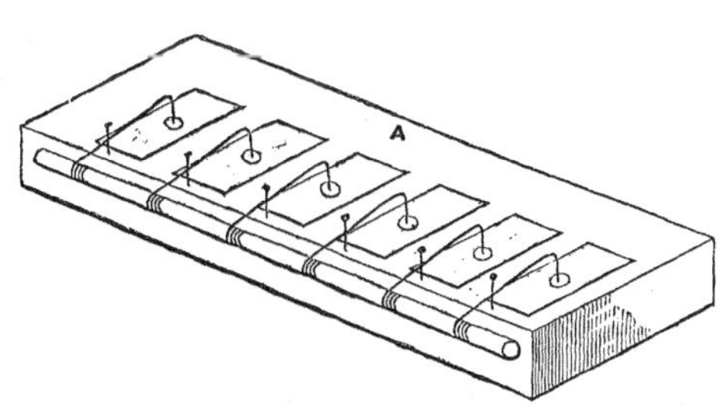

WIRE CLIP.

Mr. J. B. Spencer's model is made thus:—It is formed of thin sheet steel (obtainable at any instrument maker's), and cut out in one piece, of the form above, with a stout pair of scissors, and then bent the required shape with a pair of pliers. When used, the fore and middle fingers are applied on the under side, and the thumb on the spring. If great pressure is required, two clips may be used,—one at each end of the slide,—and for any delicate work the *width* of the steel can be reduced.

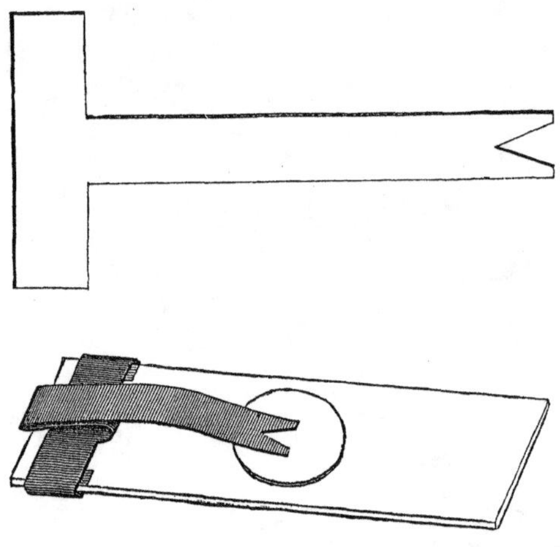

STEEL CLIP.

The American wooden spring clips are occasionally very useful, and wire clips of the kind described by Dr. Carpenter are now commonly sold and are indispensable.

Common watch-glasses should always be kept at hand. They are certainly the cheapest, and their transparency makes them very convenient reservoirs in which objects may be steeped in any liquid; and the use of them saves much trouble in examining cursorily under the microscope, whether the air-bubbles are expelled from insects, &c. &c. They are readily cleaned, and serve very well as covers, when turned upside down, to protect objects from dust. For this latter purpose Dr. Carpenter recommends the use of a number of bell-glasses, especially when one object must be left for a time (which often happens) in order that another may be proceeded with. Wine-glasses, when the legs are broken, may thus be rendered very useful.

As heat is necessary in mounting many obejects, a lamp will be required. Where gas is used, the small lamp known as "Bunsen's" is the most convenient and inexpensive. It gives great heat, is free from smoke, and is readily affixed to the common gas-burner by a few feet of india-rubber tubing. The *light* from these lamps is small, but this is little or no drawback to their use. Where gas is not available, the common spirit-lamps may be used, as they are very clean and answer every purpose.

In applying the required heat to the slides, covers, &c., it is necessary in all cases to ensure uniformity, otherwise there is danger of the glass being broken. For this purpose a brass plate at least three inches wide, somewhat longer, and one-eighth of an inch thick must be procured. It should then be affixed to a stand, so that it may be readily moved higher or lower, in order that the distance from the lamp may be changed at will, and thus the degree of heat more easily regulated. This has also the advantage of enabling the operator to allow his slides, &c., to cool more gradually, which, in some cases, is absolutely necessary,—as in fusing some of the salts, &c.

In order to get rid of air-bubbles, which are frequently disagreeable enemies to the mounter of objects, an air-pump is often very useful. This is made by covering a circular plate of metal with a bell-glass, both of which are ground so finely at the edges that greasing the place of contact renders it air-tight. The pump is then joined to the metal plate underneath, and worked with a small handle like a common syringe. By turning a small milled head the air may be allowed to re-enter when it is required to remove the bell-glass and examine or perform any operation upon the object. The mode of using this instrument will be described hereafter, but it may be here stated that substitutes have been devised for this useful apparatus; but as it is now to be obtained at a low cost, it is hardly worth while to consider them. Much time is, in many instances, certainly saved by its use, as a very long immersion in the liquids would be required to expel the bubbles, where the air-pump would remove them in an hour.

The next thing to be considered is what may be termed CEMENTS, some of which are necessary in every method of mounting objects for the microscope. Of these will be given the composition where it is probable the young student can use it; but many of them are so universally kept as to be obtainable almost anywhere; and when small quantities only are required, economy suffers more from home manufacture than from paying the maker's profit.

Amongst these, CANADA BALSAM may, perhaps, be termed the most necessary, as it is generally used for the preservation of many transparent objects. It is a thick liquid resin of a light amber-colour, which on exposure to the atmosphere becomes dry and hard even to brittleness. For this reason it is seldom used as a *cement* alone where the surface of contact is small, as it would be apt to be displaced by any sudden shock, especially when old. In the ordinary method of using, however, it serves the double purpose of preserving the object and fixing the thin glass cover; whilst the comparatively large space upon which it lies lessens the risk of displacement. By keeping, this substance becomes thicker; but a very little warmth will render it liquid enough to use, even when to some extent this change has taken place. When heated, however, for some time and allowed to cool, it becomes hardened to any degree, which may be readily regulated by the length of time it has been exposed, and the amount of heat to which it has been subjected. On account of this property it is often used with chloroform: the balsam is exposed to heat until, on cooling, it assumes a glassy appearance. This will be most readily done by baking it in what we should call a "cool oven." The time required will most likely be 20 or 30 hours. Care must be taken that the heat is not too great, else the balsam will be discoloured. It must then be dissolved in pure chloroform or benzole (the latter is preferable) until it becomes of the consistence of thick varnish. This liquid is very convenient in some cases, as air-bubbles are much more easily dispelled than when undiluted Canada balsam is used. It also dries readily, as the chloroform evaporates very quickly, for which reason it must be preserved in a closely-stoppered bottle. It has been said that this mixture becomes *cloudy* with long keeping, but I have not found it so in any cases where I have used it. Cloudiness is most frequently, if not always, caused by dampness in the object, as mentioned in Chapter IV. Should it, however, become so, a little heat will generally dispel the opacity. The ordinary balsam, if exposed much to the air whilst being used, becomes thicker, as has been already stated. It may be reduced to the required consistency with common turpentine; but I have often found this in some degree injurious to the transparency of the balsam, and the amalgamation of the two by no means perfect. (See also Chapter IV.) Its cheapness renders it no extravagance to use it always undiluted; and when preserved in a bottle with a hollow cover fitting tightly around the neck, both surfaces being finely ground, it remains fit for use much longer than in the ordinary jar. Canada balsam may now be procured in collapsible tin tubes, like those used by artists; and its manipulation is thus rendered much more easy, cleanly, and convenient, as well as economical. Chloroform is, however, frequently used for dilution, and is perhaps the safest solvent we can employ.

Dammar Varnish.—Some complain that this varnish is not easily procurable in a pure transparent state. It is often used by our American friends in mounting diatoms and other fine work. It is very liquid, and is thought by some to be more easily worked than Canada balsam. Dammar may be easily dissolved in benzole to any extent. The lumps should previously be scraped until they are freed from dust and other impurities, and then roughly crushed.

ASPHALTUM.—This substance is dissolved in linseed oil, turpentine, or naphtha, and is often termed "Brunswick black." It is easily worked, but is not generally deemed a *trustworthy* cement, as after a time it is readily loosened from its ground. It is, however, very useful for some purposes (such as "finishing" the slides), as it dries quickly. I shall, however, mention a modification of this cement a little farther on.

MARINE GLUE.—No cement is more useful or trustworthy for certain purposes than this. It is made in various proportions; but one really good mixture is—equal parts of india-rubber and gum shellac: these are dissolved in mineral naphtha with heat. It is, however, much better to get it from the opticians or others who keep it. It requires heat in the application, as will be explained in Chapter V.; but is soluble in few, if any of the liquids used by the microscopist, and for that reason is serviceable in the manufacture of cells, &c. Where two pieces of glass are to be firmly cemented together, it is almost always employed; and in all glass troughs, plates with ledges, &c., the beginner may find examples of its use.

GOLD SIZE.—This substance may always be procured at any colourman's shop. The process of its preparation is long and tedious. It is therefore not necessary to describe it here. Dr. Carpenter says that it is very durable, and may be used with almost any preservative liquids, as it is acted upon by very few of them, turpentine being its only true solvent. If too thin, it may be exposed for awhile to the open air, which by evaporation gradually thickens it. Care must be taken, however, not to render it too thick, as it will then be useless. A small quantity should be kept on hand, as it is much more adhesive when *old*.

GUM DAMMAR CEMENT.—An excellent cement may be

made by dissolving gum dammar in benzole, and adding about one-third of gold size: it dries very readily, and is especially useful when mounting objects in fluid, taking care that no moisture extends beyond the covering glass, which would prevent the complete adhesion of the cement. In those cases where glycerine is employed as the mounting medium, a ring of liquid glue put round the cover first, and when that is dry, a second coat of gum dammar will keep the cover very secure, and no leakage take place.

LIQUID GLUE is another of these cements, which is made by dissolving gum shellac in naphtha in such quantity that it may be of the required consistency. This cement appears to me almost worthless in ordinary work, as its adherence can never be relied upon; but it is so often used and recommended that an enumeration of cements might be deemed incomplete without it. Even when employed simply for varnishing the outside of the glass covers, for appearance's sake alone, it invariably chips. Where, however, oil is used as a preservative liquid, it serves very well to attach the thin glass; but when this is accomplished, another varnish less liable to chip must always be laid upon it. (See Chapter V.) Yet it makes excellent cells.

BLACK JAPAN.—This is prepared from oil of turpentine, linseed oil, amber, gum anime, and asphalt. It is troublesome to make, and therefore it is much better to procure it at the shops. It is a really good cement, and serves very well to make shallow cells for liquids, as will be described in Chapter IV. The finished cell should be exposed for a short time to the heat of what is usually termed a "cool oven." This renders it very durable, and many very careful manipulators use it for their preparations.

ELECTRICAL CEMENT.—This will be found very good for some purposes hereinafter described. To make it, melt together—

5 parts of resin.
1 „ beeswax.
1 „ red ochre.

It must be used whilst hot, and as long as it retains even slight warmth can be readily moulded into any form. It is often employed in making shallow cells for liquids, as before mentioned.

GUM-WATER is an article which nobody should ever be without; but labels, or indeed any substance, affixed to glass with common gum, are so liable to leave it spontaneously, especially when kept very dry, that I have lately added five or six drops of glycerine to an ounce of the gum solution. This addition has rendered it very trustworthy even on glass, and now I never use it without. Ten grains of moist sugar to each ounce of gum solution will also answer equally well. This solution cannot be kept long without undergoing fermentation, to prevent which the addition of a small quantity of any essential oil (as oil of cloves, &c.), or one-fourth of its volume of alcohol, may be made, which will not interfere in any way with its use.

There is what is sometimes termed an *extra adhesive* gum-water, which is made with the addition of isinglass, thus:—Dissolve two drachms of isinglass in four ounces of distilled vinegar; add as much gum arabic as will give it the required consistency. This will keep very well, but is apt to become thinner, when a little more gum may be added.

I may here mention that Messrs. Marion have lately brought out a cement for the purpose of mounting photographs, which is very adhesive, even to glass. I find it useful in all cases where certainty is requisite; as gummed paper is liable in a dry place to curl from the slides, as before mentioned.

All these, except one or two, are liquid, and must be kept in *stoppered* bottles, or, at least, as free from the action of the air as possible.

When any two substances are to be united firmly, I have termed the medium employed "a cement;" but often the appearance of the slides is thought to be improved by drawing a coloured ring upon them, extending partly on the cover and partly on the slide, hiding the junction of the two. The medium used in these cases I term A VARNISH, and hereinafter mention one or two. Of course, the tenacity is not required to be so perfect as in the *cements*.

SEALING-WAX VARNISH is prepared by coarsely powdering sealing-wax, and adding spirits of wine; it is then digested at a gentle heat to the required thickness. This is very frequently used to finish the slides, as before mentioned, and can easily be made of any colour by employing different kinds of sealing-wax; but is very liable to chip and leave the glass. The best qualities, however, will be less liable.

BLACK VARNISH—Is readily prepared by adding a small quantity of lampblack to gold-size and mixing intimately. Dr. Carpenter recommends this as a good finishing varnish, drying quickly and being free from that brittleness which renders some of the others almost worthless; but it should not be used in the first process when mounting objects in fluid.

Amongst these different cements and varnishes I worked a long time without coming to any decision as to their comparative qualities, though making innumerable experiments. The harder kinds were continually cracking, and the softer possessed but little adhesive power. To find hardness and adhesiveness united was my object, and the following possesses these qualities in a great degree:—

India-rubber	½ drachm.
Asphaltum	4 oz.
Mineral naphtha	10 „

Dissolve the india-rubber in the naphtha, then add the asphaltum—if necessary, heat must be employed.

Some scientific friends have complained that they have been unable to dissolve either the india-rubber or the asphaltum in mineral naphtha. The frequency with which I have seen this solution thoroughly accomplished convinces me that one of these things has occurred—either the india-rubber or the asphaltum has not been pure, or the naphtha has been *wood* instead of *mineral*. In the early photographic days every artist made a form of this varnish to use with glass positives, and I never heard a complaint of difficulty.

This is often used by photographers as a black varnish for glass, and never cracks, whilst it is very adhesive. Dr. Carpenter, however, states that his experience has not been favourable to it; but I have used it in great quantities and have never found it to leave the glass in a single instance when used in the above proportions. The objections to it are, however, I think easily explained, when it is known that there are many kinds of pitch, &c., from coal, sold by the name of asphaltum, some of which are worthless in making a microscopic cement. When used for this purpose, the asphaltum must be genuine and of the best quality that can be bought. The above mixture serves a double purpose—to unite the cell to the slide, and also as a "finishing" varnish. But it is perhaps more convenient to have two bottles of this cement, one of which is thicker than common varnish, to use for uniting the cell, &c.; the other liquid enough to flow readily, which may be employed as a surface varnish in finishing the slides.

The brushes or camel-hair pencils should always be cleaned after use; but with the asphalt varnish above mentioned it is sufficient to wipe off as carefully as possible the superfluous quantity which adheres to the pencil, as, when again used, the varnish will readily soften it; but, of course, it will be necessary to keep separate brushes for certain purposes.

Here it may be observed that every object should be labelled with name and any other descriptive item as soon as mounted. There are many little differences in the methods of doing this. Some write with a diamond upon the slide itself; but this has the disadvantage of being not so easily seen. For this reason a small piece of paper is usually affixed to one end of the slide, on which is written what is required. These labels may be bought of different colours and designs; but the most simple are quite as good, and very readily procured. Take a sheet of thin writing paper and brush over one side a strong solution of gum, with the addition of a few drops of glycerine, or grains of moist sugar, as above recommended; allow this to dry, and then with a common gun-punch stamp out the circles, which may be affixed to the slides by simply damping the gummed surface, taking care to write the required name, &c., upon it before damping it, or else allowing it to become perfectly dry first.

There is one difficulty which a beginner often experiences in sorting and mounting certain specimens under the microscope, viz., the *inversion* of the objects; and it is often stated to be almost impossible to work without an erector. But this difficulty soon vanishes, the young student becoming used to working what at first seems in contradiction to his sight.

Let it be understood, that in giving the description of those articles which are usually esteemed *necessary* in the various parts of microscopic manipulation, I do not mean to say that without many of these no work of any value can be done. There are, as all will allow, certain forms of apparatus which aid the operator considerably; but the cost may be too great for him. A little thought, however, will frequently overcome this difficulty, by enabling him to make, or get made, for himself, at a comparatively light expense, something which will accomplish all he desires. As an example of this, a friend of mine made what he terms his "universal stand," to carry various condensers, &c. &c., in the following way:—Take a steel or brass wire, three-sixteenths or one-quarter inch thick and six or eight inches long; "tap" into a *solid*, or make rough and fasten with melted lead into a *hollow*, ball. (The foot of a cabinet or work-box answers the purpose very well.) In the centre of a round piece of tough board, three inches in diameter, make a hemispherical cavity to fit half of the ball, and bore a hole through from the middle of this cavity, to allow the wire to pass. Take another piece of board, about four inches in diameter, either round or square, and one and a half or two inches thick, make a similar cavity in its centre to receive the other half of the ball, but only so deep as to allow the ball to fit tightly when the two pieces of board are screwed together, which last operation must be done with three or four screws. Let the hole for the wire in the upper part be made conical (base upwards), and so large as only to prevent the ball from escaping from its socket, in order that the shaft may move about as freely as possible. Turn a cavity, or make holes, in the bottom of the under piece, and fill with lead to give weight and steadiness. This, painted green bronze and varnished, looks neat; and by having pieces of gutta-percha tubing to fit the shaft, a great variety of apparatus may be attached to it.

Mr. Loy employs the following arrangement for dissecting insects or picking out Foraminifera, &c.: he fits an upright brass rod into a heavy leaden foot, this rod carries a horizontal arm bearing at its end a ring for holding a watchmaker's eye-glass; in focussing it to his work, he presses the eye-glass down with his head, the weight of the leaden foot keeping it in its place, and allowing it to follow his every movement.

Again, a "condenser" is often required for the illumination of opaque objects. My ingenious friend uses an "engraver's bottle" (price 6d.), fills it with water, and suspends it betwixt the light and the object. Where the light is very yellow, he tints the water with indigo, and so removes the objectionable colour.

I merely mention these as examples of what may be done by a little thoughtful contrivance, and to remove the idea that nothing is of much value save that which is the work of professional workmen, and consequently expensive.

CHAPTER III.

TO PREPARE AND MOUNT OBJECTS "DRY."

THE term "dry" is used when the object to be mounted is not immersed in any liquid or medium, but preserved in its natural state, unless it requires cleaning and drying.

I have before stated that thorough cleanliness is necessary in the mounting of all microscopic objects. I may here add that almost every kind of substance used by the microscopist suffers from careless handling. Many leaves with fine hairs are robbed of half their beauty, or the hairs, perhaps, forced into totally different shapes and groups; many insects lose their scales, which constitute their chief value to the microscopist; even glass itself distinctly shows the marks of the fingers if left uncleaned. Every object must also be *thoroughly dry*, otherwise dampness will arise and become condensed in small drops upon the inner surface of the thin glass cover. This defect is frequently met with in slides which have been mounted quickly; the objects not being thoroughly dry when enclosed in the cell. Many cheap slides are thus rendered worthless. Even with every care it is not possible to get rid of this annoyance occasionally. A good plan is to fix the covers on to the cells temporarily by dropping on two sides of them a composition of equal parts of wax and resin: this allows of the easy removal of the cover at any time, while the object thoroughly dries and is protected from dust and damage.

For the purpose of mounting opaque objects "dry" *discs* were at one time very commonly used. These are circular pieces of cork, leather, or other soft substance, from one-quarter to half an inch in diameter, blackened with varnish or covered with black paper, on which the object is fixed by gum or some other adhesive substance. They are usually pierced longitudinally by a strong pin, which serves for the forceps to lay hold of when being placed under the microscope for examination. Sometimes objects are affixed to both sides of the disc, which is readily turned when under the object-glass. The advantage of this method of mounting is the ease with which the disc may be moved, and so present every part of the object to the eye, save that by which it is fastened to the disc. On this account it is often used when some particular subject is undergoing investigation, as a number of specimens may be placed upon the discs with very little labour, displaying all their parts. But where exposure to the atmosphere or small particles of dust will injure an object, no advantage which discs may possess should be considered, and an ordinary covered cell should be employed. Small pill-boxes have been used, to the bottom of which a piece of cork has been glued to afford a ground for the pin or other mode of attachment; but this is liable to *some* of the same faults as the disc, and it would be unwise to use these for permanent objects.

Messrs. Smith and Beck have lately invented, and are now making a beautiful small apparatus, by means of which the disc supporting the object can be worked with little or no trouble into any position that may prove most convenient, whilst a perforated cylinder serves for the reception of the discs when out of use, and fits into a case to protect them from dust. A pair of forceps is made for the express purpose of removing them from the case and placing them in the holder.

All dry objects, however, which are to be preserved should be mounted on glass slides in one of the cells (described in Chapter II.) best suited to them. Where the object is to be free from pressure, care must be taken that the cell is deep enough to ensure this. When the depth required is but small, it is often sufficient to omit the card, leather, or other circles, and with the "turntable" before described by means of a thick varnish and camel-hair pencil, to form a ring of the desired depth; but should the varnish not be of sufficient substance to give such "walls" at once, the first application may be allowed to dry, and a second made upon it. A number of these may be prepared at the same time, and laid by for use. When liquids are used (see Chapter V.), Dr. Carpenter recommends gold-size as a good varnish for the purpose, and this may be used in dry mountings also. I have used the asphaltum and india-rubber (mentioned in Chapter II.), and found it to be everything I could wish. The cells, however, must be *thoroughly dry*, and when they will bear the heat they should be baked for an hour at least in a tolerably cool oven, by which treatment the latter becomes an excellent medium. All dry objects which will not bear pressure must be firmly fastened to the slide, otherwise the necessary movements often injure them, by destroying the fine hairs, &c. For this purpose thin varnishes are often used, and will serve well enough for large objects, but many smaller ones are lost by adopting this plan, as for a time, which may be deemed long enough to harden the varnish, they exhibit no defect, but in a while a "wall" of the plastic gum gathers around them, which refracts the light, and thus leads the student to false conclusions. In all *finer* work, where it is necessary to use any method of fixing them to the slide, a solution of common gum, with the addition of a few drops of glycerine (Chapter II.), will be found to serve the purpose perfectly. It must, however, be carefully filtered through blotting-paper, otherwise the minute particles in the solution interfere with the object, giving the slide a dusty appearance when under the microscope.

When mounting an object in any of these cells, the glass must be thoroughly cleaned, which may be done with a cambric handkerchief, after the washing mentioned in Chapter II. *If the object be large*, the point of a fine camel-hair pencil should be dipped into the gum solution, and a minute quantity of the liquid deposited in the cell where the object is to be placed, but not to cover a greater surface than the object will totally hide from sight. This drop of gum must be allowed to dry, which will take a few minutes. Breathe then upon it two or three times, holding the slide not far from the mouth, which will render the surface adhesive. Then draw a camel-hair pencil through the lips, so as to moisten it slightly (when anything small will adhere to it quite firmly enough), touch the object and place it upon the gum in the desired position. This must be done immediately to ensure perfect stability, otherwise the gum will become at least partially dry and only retain the object imperfectly.

When, however, the objects are so minute that it would be impossible to deposit atoms of gum small enough for each one to cover, a different method of proceeding must be adopted. In this case a small portion of the same gum solution should be placed upon the slide, and by means of any small instrument—a long needle will serve the purpose very well—spread over the surface which will be required. The quantity thus extended will be very small, but by breathing upon it may be prevented drying whilst being dispersed. This, like the forementioned, should be then allowed to dry; and whilst the objects are being placed on the prepared surface, breathing upon it as before will restore the power of adherence. A small patch of gold-size—or gum dammar solution which has been allowed to become "tackey"—is very useful in many cases.

When gum or other liquid cement has been used to fix the objects to the glass, the thin covers must not be applied until the slide has been *thoroughly dried*, and all fear of dampness arising from the use of the solution done away with. Warmth may be safely applied for the purpose, as objects fastened by this method are seldom, if ever found to be loosened by it. As objects are met with of every thickness, the cells will be required of different depths. There is no difficulty in accommodating ourselves in this—the deeper cells may be readily cut out of thick leather, card, or other substance preferred (as mentioned in Chapter II.). Cardboard is easily procured of almost any thickness; but sometimes it is convenient to find a thinner substance even than this. When thin glass is laid upon a drop of any liquid upon a slide, every one must have observed how readily the liquid spreads betwixt the two: just so when any thin varnish is used to surround an object of little substance, excessive care is needed lest the varnish should extend betwixt the cover and slide, and so render it worthless. The slightest wall, however, prevents this from taking place, so that a ring of common paper may be used, and serve a double purpose where the objects require no deeper cell than this forms.

Many objects, however, are of such tenuity—as the leaves of many mosses, some of the Diatomaceæ, scales of insects, &c.—that no cell is requisite excepting that which is necessarily formed by the medium used to attach the thin glass cover to the slide; and where the slide is covered by the ornamental papers mentioned in Chapter II., and pressure does not injure the object, even this is omitted, the thin glass being kept in position by the cover; but slides mounted in this manner are frequently injured by dampness, which soon condenses upon the inner surfaces and interferes both with the object and the clearness of its appearance.

The thin glass, then, is to be united to the slide, so as to form a perfect protection from dust, dampness, or other injurious matter, and yet allow a thoroughly distinct view of the object. This is to be done by applying to the glass slide round the object some adhesive substance, and with the forceps placing the thin glass cover (quite dry and clean) upon it. A gentle pressure round the edge will then ensure a perfect adhesion, and with ordinary care there will be little or no danger of breakage. For this purpose gold-size is frequently used. The asphalt and india-rubber varnish also will be found both durable and serviceable. Whatever cement may be used, it is well to allow it to become in some measure fixed and dried; but where no cell or wall is upon the slide, this is *quite necessary*, otherwise the varnish will be most certain to extend, as before mentioned, and ruin the object. It may be stated here that gold-size differs greatly in its drying powers, according to its age, mode of preparation, &c. (Chapter V.): here gum dammar solution laid on in a very thin coating will be found most useful, as it dries so rapidly that it cannot run in unless laid on with an unsparing hand.

Should any object be enclosed which requires to be kept flat during the drying of the cement, it will be necessary to use some of the contrivances mentioned in Chapter II.

When the slide is thus far advanced, there remains the finishing only. Should the student, however, have no time to complete his work at once, he may safely leave it at this stage until he have a number of slides which he may finish at the same time. There are different methods of doing this, some of which may be here described.

If ornamental papers are preferred, a small circle must be cut out from the centre a little less than the thin glass which covers the object. Another piece of coloured paper is made of the same size, and a similar circle taken from *its* centre also, or both may be cut at the same time. The slide is then covered round the edges with paper of any plain colour, so that it may extend about one-eighth of an inch over the glass on every side. The ornamental paper is then pasted on the "object" surface of the glass, so that the circle shows the object as nearly in the centre as possible, and covers the edges of the thin glass. The other coloured paper is then affixed underneath with the circle coinciding with that above. And here I may observe, that when this method is used there is no necessity for the edges of the slide to be ground, as all danger of scratching, &c., is obviated by the paper cover.

Many now use paper covers, about one and a half inch long, on the upper side of the slide only, with the centre cut out as before, with no other purpose than that of hiding the edge of the thin glass where it is united to the slide.

The method of finishing, however, which is mostly used at the present time, is to lay a coating of varnish upon the edge of the thin glass, and extend it some little way on the slide. When a black circle is required, nothing serves the purpose better than the gold-size and lampblack, or the asphalt and india-rubber varnish, neither of which is liable to chip; but when used for this, the latter should be rather thinner, as before advised. Some of these varnishes are preferred of different colours, which may be made by using the different kinds of sealing-wax, as described in Chapter II.; but they are always liable to the defects there mentioned. This circle cannot be made in any other way than by one of those contrivances called turntables. A very little practice will enable the young student to place his slide so that the circle may be uniform with the edge of the thin glass.

The slide is now complete, except the addition of the name and any other particulars which may be desirable. For this purpose one of the methods described in Chapter II. must be employed.

Amongst the various classes of microscopic objects now receiving general attention, the Diatomaceæ may be placed in a prominent position. They afford endless opportunities of research, and some very elaborate works have already been issued concerning them. Professor Smith's may be mentioned as one containing, perhaps, the best illustrations. The young student may wish to know what a diatom is. The "Micrographic Dictionary" gives the following definition:—"A family of confervoid Algæ, of very peculiar character, consisting of microscopic brittle organisms." They are now looked upon by almost all of our scientific men as belonging to the *vegetable* kingdom, though some few still assign them to the animal. They are almost invariably so exceedingly small, that the unaided eye can perceive nothing on a prepared slide of these organisms but minute dust. Each separate portion, which is usually seen when mounted, is termed a "frustule," or "testule:" this consists of two similar parts, composed of silica, between and sometimes around which, is a mass of viscid matter called the "endochrome." They are found in almost every description of water, according to the variety: some prefer sea-water, others fresh, and many are seen nowhere but in that which is a mixture of both, as the mouths of rivers, &c. Ditches, ponds, cisterns, and indeed almost every *reservoir*, yield abundance of these forms. They are not, however, confined to "present" life; but, owing to the almost indestructible nature of their siliceous covering, they are found in a fossil state in certain earths in great abundance, and are often termed "fossil Infusoria." Upon these frustules are generally to be seen lines, or markings, of different degrees of minuteness, the delicacy of which often serves the purpose of testing the defining power of object-glasses. Some of the frustules are triangular, others circular, and, indeed, of almost every conceivable shape, many of them presenting us with exquisitely beautiful designs.

The markings, however, are seldom seen well, if at all, until the frustules are properly prepared, the different methods of accomplishing which will be given a little further on.

The *collection* of fresh diatoms is so closely connected with their *preservation*, that a few notes may be given upon it before we pass on. For this purpose a number of small bottles must be provided, which may be placed in a tin box, with a separate compartment for each, so that all chance of breakage may be done away with. The diatoms are generally of a light brown colour; and where they are observed in the water, the bottle may be so placed, with the mouth closed by the finger, that when the finger is withdrawn the water will rush in, carrying the diatoms also. If they are seen upon plants, stones, or any other substance, they may generally be detached and placed in the bottle. When there is a green covering upon the surface of the water, a great quantity of diatoms is usually found amongst it; as also upon the surface of the mud in those ponds where they abound. In these cases, a broad flat spoon will be found very useful, and one is now made with a covering upon the broader portion of it to protect the enclosed matter from being so readily carried off whilst bringing it to the surface again. Where there is any depth of water, and the spoon will not reach the surface of the mud, the bottle must be united to a long rod, and being then carried through the upper portion with the mouth downwards, no water will be received into it; but on reaching the spot required, the bottle-mouth may be turned up, and thus become filled with what is nearest.

From the stomachs of common fish—as the cod, sole, haddock, &c.—many specimens of Diatomaceæ may be obtained, but especially from the crab, oyster, mussel, and other shell-fish. Professor Smith states that from these curious receptacles he has taken some with which he has not elsewhere met. To remove them from any of the small shell-fish, it is necessary to take the fish or stomach from the shell, and immerse it in strong hot acid (nitric is the best) until the animal matter is dissolved, when the residue must be washed and treated as the ordinary Diatomaceæ hereinafter described.

Many diatoms are seen best when mounted in a dry state, the minute markings becoming much more indistinct if immersed in liquid or balsam; and for this reason those which are used as test objects are usually mounted *dry*. Many kinds are also now prepared in this way as opaque objects, to be examined with the lieberkuhn, and are exquisitely beautiful. Others, however, are almost invariably mounted in balsam; but as these will be again referred to in Chapter IV., and require the same treatment to fit them for the slide, it will not be out of place to describe the cleaning and preparation of them here. As before stated, there is much matter surrounding them which must be got rid of before the *siliceous* covering can be shown perfectly. As, however, we may first wish to become acquainted in some degree with what we have to do, it is well to take a small piece of *talc*, and place a few of the diatoms upon it. This may be held over the flame of the spirit-lamp until all the surrounding matter is burnt away, and a tolerable idea may thus be obtained as to the quality of our treasure.

In some cases it is well to use this burning operation alone in *mounting* specimens of diatoms, when they may be placed in their natural state upon thin glass, burnt for awhile upon the platinum plate, hereinafter described, and mounted dry or in balsam.

In the preparation and cleaning of Diatomaceæ, there is little satisfaction unless these operations have been successfully performed, as a very small portion of foreign matter seriously interferes with the object. The mode of preparing them varies even amongst the most experienced. It will be found, therefore, most satisfactory to examine the principal of these separately, although it may be at the risk of some little repetition.

The method which is most frequently employed is the following:—Place the gathering containing the Diatomaceæ in a small glass or porcelain vessel, add strong nitric acid, and, by the aid of a Bunsen's burner or spirit-lamp, boil for some minutes. From time to time a drop of the mixture may be put upon a slide, and examined under the microscope to see if all foreign matter be got rid of. When the valves are clean, the vessel containing them must be filled with water, and the whole left for an hour or two, so that all the diatoms may settle perfectly. The liquid must then be poured off carefully, or drained away by the aid of a syphon, so that none of the diatoms are removed with it. Indeed, it is well to examine the liquid drained off each time with the microscope, as the finer forms are frequently lost in the washings. The vessel must then be refilled with pure water, allowed to settle, and drained as before. This washing must be repeated until a drop being placed upon a slide and evaporated leaves no crystals. When it is desirable to preserve the diatoms in this state before mounting (which process will be described in another place), they may be placed in a small phial with a little distilled water.

There are many cases in which the above method will not effect a *perfect* cleansing, as certain substances with which diatoms are frequently mixed are not soluble in *nitric acid.* For this reason the following method is resorted to:—Take a quantity of the matter containing the Diatomaceæ and wash first with pure water, to get rid of all the impurities possible. Allow this to settle perfectly and decant the water. Add hydrochloric acid gradually, and when all effervescence has subsided, boil for some minutes by aid of the lamp. When cool and the particles have subsided, decant the hydrochloric and add nitric acid. The boiling must then be repeated until a drop of the liquid when placed under the microscope shows the valves or "frustules" clean. After allowing the diatoms to settle, the acid must be decanted, and pure water substituted. The washing must be repeated as in the former process until all the remains of crystals or acid are removed, when the specimens may be preserved in small phials.

Such are the usual modes of treating the Diatomaceæ, but there are certain cases in which particular methods are required to give anything like perfect results. Persons of great experience combine a variety of treatments, and thus obtain better and more uniform specimens. Perhaps it will be advantageous to give the young student the process adopted by one of the most successful preparers of these objects; but I will first state the different methods of mounting the cleaned diatoms dry: how to employ Canada balsam and fluid in their preservation will be elsewhere described.

It was before stated that diatoms when cleansed might be preserved in small phials of distilled water. When required for mounting, shake the phial, and with a thin glass tube or rod take up a drop of the fluid and spread it upon the surface of the slide in the desired position. This must then be allowed to dry gradually, or by the aid of the lamp if necessary, without being shaken or interfered with, otherwise uniformity of dispersion will be prevented. When *thoroughly* dry, a thin ring of one of the adhesive varnishes —gold-size will be found as good as any—may be drawn round the diatoms, and allowed to dry in a slight degree. The slide and thin glass cover should then be warmed and the latter gently pressed upon the ring of varnish until the adhesion all round is complete.

As some of the diatoms require object-glasses of extremely high power, and consequently short focus, to show them, they must be as close to the outer surface of the cover as possible. For this reason they are sometimes placed upon the under side of the thin glass, as follows. Clean the surfaces of the slide and cover, and with the rod or pipe place the liquid containing the diatoms upon the thin glass, and dry as before. Trace the ring to receive the cover upon the slide, and when almost dry, warm both and proceed as above. Whichever of these methods is employed, the outer ring of coloured varnish may be applied as elsewhere described and the slide finished.

Diatoms are also sometimes mounted betwixt two thin glasses, as described in Chapter II., so that the light by which they are examined may receive as little interference as possible, and that an achromatic condenser may be brought into focus under the slide.

Of the various modes of cleaning and mounting Diatomaceæ, I believe that the following may be safely recommended, as affordiug results of the best quality. My friend, Mr. T. G. Rylands, gave it to me as that which he prefers, and I can safely say that his numerous slides are at least equal to any I have ever seen. I will give it just as I received it from him, though there may be some little repetition of what has been said elsewhere, as he does not appropriate any part of it as his own. He says:—In this branch of mounting, general rules alone can be laid down, because the gatherings may contain iron, lime, fine silt, or vegetable matter under conditions for special treatment, and consequently the first step should be to experiment on various kinds.

In gathering diatoms much labour is saved by judgment and care; hence it is desirable to get acquainted with them in their growing condition, so that when recognised upon the sand or other spots, they may be carefully removed by the aid of the spoon or small tin scoop before described. When growing upon algæ or other plants, the plants and diatoms together may be carried home, in which case they must be simply drained and not washed or pressed, in order that the diatoms be not lost. As it is always desirable to examine the gathering on the ground, a "Gairdner's hand microscope" with powers from 80 to 200 diameters will be found very useful. The best gatherings are those which represent one species abundantly. Those which are mixed may be rejected, unless they are seen to contain something valuable or important, as the object should be not so much to supply microscopical curiosities as to collect material which is available for the study of nature.

The gathering when carried home should always be carefully examined before anything is done with it; not only on account of the additional information thus acquired, but also because it often happens that a specimen should be mounted in fluid (see Chapter V.) in the condition in which it is gathered, as well as cleaned and mounted in balsam (Chapter IV.) and dry.

Where the gathering is taken from sand, the whole may be shaken up in water as a preliminary operation, when much of the sand will be separated by its own weight. The lime test, however, should be applied; viz.—a small portion of hydrochloric acid, and if there be effervescence, it must be dissolved out by this means. From Algæ and other weeds diatoms may be detached by agitating the whole together in a weak solution of nitric acid—about one of pure acid to twenty or thirty of water, as it must be sufficiently weak to free the diatoms without destroying the matter to which they adhere. The diatoms may then be separated by sifting through coarse muslin, which will retain the Algæ, &c. The process of cleaning will vary according to circumstances. Some gatherings require to be boiled only a few minutes in nitric acid; but the more general plan, where they are mixed with organic or other foreign matter, is to boil them in pure sulphuric acid until they cease to grow darker in colour (usually from a half to one minute), and then to add, drop by drop to avoid explosions, a cold saturated solution of chlorate of potash until the colour is discharged, or, in case the colour does not disappear, the quantity of the solution used is at least equal to that of the acid. This operation is best performed in a wide-mouthed ordinary beaker glass,* a test-tube being too narrow. The mixture whilst boiling should be poured into thirty times its bulk of cold water, and the whole allowed to subside. The fluid must then be carefully decanted and the vessel re-supplied once or twice with pure water, so as to get rid of all the acid. The gathering may then be transferred to a small boiling-glass or test-tube, and—the water being carefully decanted—boiled in the smallest available quantity of nitric acid, and washed as before. This last process has been found necessary from the frequent appearance of minute crystals, which cannot otherwise be readily disposed of without the loss of a considerable proportion of diatoms.

I may here mention that the washing-glasses used by Mr. Rylands are stoppered conical bottles varying in capacity from two ounces to one quart; the *conical* form being employed to prevent the adherence of anything to the side: they are stoppered, to render them available in the shaking process about to be described.

The gathering, freed from acid, is now put into two inches depth of water, shaken vigorously for a minute or two, and allowed to subside for half an hour, after which the turbid

* These glasses are round, about six inches high, and usually contain about eight ounces. They are rather wider at the bottom, tapering gradually to the top, and may be generally procured at the chemists, &c.

fluid must be carefully decanted. This operation must be repeated until all the matter is removed which will not settle in half an hour. The fluid removed should be examined by a drop being put upon a slide, as in some cases very light diatoms have been found to come off almost pure in one or more of these earlier washings. The quantity of water and time of subsidence given may be taken generally, but may require to be modified according to circumstances and the judgment of the operator. By the repetition and variation of this process—the *shaking* being the most important part—the gathering, if a *pure* one, will be sufficiently clean. If, however, it contain a variety of species and forms, it may require to be divided into *different densities.*

In some cases, however, it is best to divide the gathering as a *preliminary* operation, which may be done by agitating it in a quantity of water and decanting what does not readily subside. The heavier and the lighter portions are then to be treated as two separate boilings. But when the cleansing has been carried to the above stage and this division is required, the plan must be somewhat as follows:—The gathering must be shaken in a test-tube with six inches of water, and then allowed to subside until one inch at the top remains pure. About three inches are then to be carefully withdrawn by a pipette, when the tube may be filled up and the operation repeated. The three lower inches also may then be decanted and examined. The gathering is thus divided into three portions; viz.—that which was withdrawn by the pipette, that which remained floating in the lower three inches of water in the tube, and that which had settled at the bottom. An examination of these will inform the operator how to obtain that particular density of gathering which he desires, and how far it is worth while to refine this process of elutriation; for in cases of necessity any one, or all three, of these densities may be operated upon in the same way to separate a particular diatom.

As an occasional aid, it may be remarked, that in some cases *liquor ammoniæ* may be used in one of the later washings in place of water, as it often separates fine dirt, which is not otherwise easily removed. Ammonia also dissolves a flocculent matter which sometimes remains; and this method does not injure diatoms like some strong alkalies. Some fossil deposits require to be treated with a boiling solution of carbonate of soda to disintegrate them; but this operation requires great care, lest the alkali should destroy the diatoms. Vegetable silicates also sometimes require to be removed by a solution of carbonate of soda; but as the frustules of the diatoms themselves are but *vegetable silica,* even more care is required in this case. It may be well to mention, that some diatoms are so imperfectly siliceous that they will not bear *boiling* in acid at all. Some of these may be allowed to stand in cold nitric acid some time, whilst others of a smaller and more delicate character should, when possible, be treated with distilled water alone.

We will now consider the mode of mounting the prepared diatoms, which, if used dry (as described in this chapter), should be carefully washed two or three times with the purest distilled water. In this branch, as in every other, each collector gives preference to that method in which he is an adept. Thus the diatoms may be placed on the under side of the cover, to be as near to the object-glass as possible, or upon the slide itself; and each plan has its advocates. Whichsoever of these is used, nothing seems more simple to the novice than a tolerably equal dispersion of the objects upon the slide or cover; but this is by no means so readily accomplished; consequently I give Mr. Rylands' method, as his slides are perfect in this respect also. He always places the diatoms upon the thin glass cover. It is not sufficient, as is frequently thought, to take a drop of liquid containing the cleansed material and spread it upon the cover or slide, as without some additional precaution that uniform and regular distribution of the specimens, which is desirable, is not obtained. In order to effect this, let a drop of the cleansed gathering be diluted sufficiently for the purpose—how much must be determined in each case by experiment—and let the covers to be mounted be cleaned and laid upon the brass plate. (See Chapter II.) By means of a glass tube, about one-twelfth of an inch in diameter, stopped by the wetted finger at the upper end, take up as much of the diluted material as will form a moderately convex drop extending over the whole cover. When all the covers required are thus prepared, apply a lamp below the brass plate, and raise the temperature to a point just short of boiling. By this means the covers will be dried in a few minutes, and the specimens equally distributed over the whole area. The spread of the fluid upon the covers is facilitated by breathing upon them; and, to insure uniformity, care must be taken to avoid shaking them whilst drying. The best plan is to mount at least half a dozen at once.

Before mounting, Mr. Rylands always burns the diatoms upon the glass at a dull red heat, whether they are used with balsam or dry. This burning, he says, is not only an additional cleaning process, but it effectually fixes the diatoms, and prevents them floating out if mounted with balsam. The thinnest covers may be burnt without damage if they are placed upon a small piece of platinum foil of the size required, which should be about one-hundredth of an inch thick, perfectly flat, and having three of its edges slightly bent over, so as to prevent its warping with the heat. The small flame of a spirit-lamp, or, where there is gas, a Bunsen's burner, may be employed. The cover should be shaded from direct daylight, that the action of the flame may be observed more perfectly. Care must then be taken to raise the temperature only to the dull red heat before mentioned. The cover will then be in a fit state for mounting as required.

It has been stated in another place that it is assumed that the operator is not mounting diatoms simply as microscopic objects, but as instructive specimens. It is not, therefore, sufficient to take a single slide as all that is required, but to have the same diatom prepared in as many ways as possible. The following are the principal:—

1. Mounted crude in fluid (see Chapter V.)
2. Burnt crude upon the cover, and mounted dry or in balsam.
3. Mounted dry or in balsam (see Chapter IV.), after the cleansing process already described.

I will here give Mr. Rylands' method of mounting them *dry*, the fluid and balsam preparations being noticed in their respective chapters. The slide, with the ring of asphalt, or black varnish, should have been prepared some weeks previously, in order to allow it to dry thoroughly. When required, it must be held over the spirit-lamp or Bunsen's burner until the ring of varnish is softened. The burnt cover, having been heated at the same time, must then be taken in the forceps and pressed upon the softened varnish until it adhere all round. When cold, an outer ring of asphalt completes the slide.

Such is the method which my friend Mr. T. G. Rylands employs in the preparation of diatoms for the microscope. I have said enough concerning his results. It is to be feared, however, that to some these several modes of operation may appear lengthy and complicated; but if read carefully, and the experiments tried, they will be found to be simple enough in practice, and to occupy much less time than an intelligible description would lead the novice to believe necessary,

The minute nature of diatom forms, and the high microscopic powers by which they are examined, render a very shallow cell necessary when they are mounted upon a dry slide. Many early attempts, on this account, have been ruined by the cement used to fix the thin cover spreading underneath. A correspondent of the "Monthly Microscopic Journal" thus gives his mode of avoiding this:—"There is a very simple means of avoiding this danger, and I will now describe it. A circle of bitumen about one-third smaller than the covering glass is drawn beforehand on my slides. When I wish to make a preparation, instead of coating, as formerly, the first circle with a second layer of bitumen, I form a second circle of it outside the first, and as near as possible to it, and each of the two circles has its own advantage; the first, in fact, while forming the cell, serves as a support for the covering glass, and thus preserves the Diatomaceæ from any breakage; it offers, besides, a serious obstacle to the spreading of the more liquid bitumen of which the outer circle is composed; and the latter closes the cell by fixing the cover, which, when the preparation is dry, may be covered with a final circle of bitumen. It is of course understood that I am speaking of preparations made in the dry way only, and not with balsam."

One of the most fertile as well as the most curious magazines of Diatomaceæ is *guano*. The siliceous forms contained therein have been devoured by sea-birds and passed through the stomach uninjured, and after lying for ages may be cleaned and classified. Many of these are not elsewhere met with, so that the student who is desirous to enter into the study of Diatomaceæ must be instructed as to the best mode of obtaining them from this source. The particulars to be observed so closely resemble those before mentioned in the treatment of the ordinary diatoms, that it will be sufficiently explicit to give the outlines of the process. The guano must be first washed in pure water, allowed to subside perfectly, and the liquid then poured off. This must be repeated until the top fluid is clear, and care taken not to decant the liquid until perfect subsidence has taken place. The deposit must then be treated with hydrochloric acid with a gentle heat for an hour or two, adding a little fresh acid at intervals as long as it excites any effervescence. After this nitric acid must be substituted for the hydrochloric, and the heat kept up to almost boiling-point for another hour at least, adding a little fresh acid as before. When this ceases to act, the deposit must be allowed to settle perfectly and the acid poured off. All traces of the acid must now be washed away with pure water, when the remains will be Diatomaceæ, the sand contained in the guano, and a few other forms. Some of these may be mounted dry, as before mentioned, but the greater portion should be put up in Canada balsam as described in Chapter IV.

Such is the ordinary method for the treatment of guano; but Mr. Rylands' mode of proceeding with ordinary Diatomaceæ (before given) will be found equally successful with these deposits.

The composition, however, of guano is more complex than the substances by which we find most of the Diatomaceæ surrounded, and therefore a different method of treatment is pursued by many. The following by Mr. A. J. Roberts is a good one: "Guano should be first well washed with boiling water, either on a paper filter or by repeated affusions until the liquid come off tasteless. Boiling water is preferable to cold, for the heat expels air-bubbles and causes the deposit to settle down into a smaller space; then the deposit must be subjected to the action of the acids as directed for the preparation of earths, to get rid of the lime salts. The partially cleaned deposit, which is now much smaller in bulk, must be separated as much as possible from the liquid, strong sulphuric acid, in sufficient quantity to cover it, poured on to about the depth of half an inch, and heat applied and continued for some time, but the liquid must not be made to boil. The result will be an almost immediate blackening of the liquid, which gradually becomes deeper, and a dirty, inky-looking compound is ultimately produced. When this has taken place, chlorate of potash in fine powder must be dropped into the hot mixture very gradually until the black colour disappear. This must be done cautiously, for the action is so violent, that much spurting is occasioned; and the liquid being very corrosive, a tolerably capacious vessel should be used in order to keep the splashes within reasonable bounds, or serious damage to the operator's clothes may ensue. The nearly decolorized liquid must now be diluted with a considerable quantity of water, and the deposit allowed to subside, the supernatant liquid poured off, and the process of heating with sulphuric acid and addition of chlorate of potash repeated until the sulphuric acid occasion no further blackening; then the cleaning may be finished in the usual way by washing."

I have had many complaints from my friends that with all their care they have found nothing fit for mounting in guano. This is readily accounted for by one fact, that not one sample in twenty of what is called guano in the market has an atom of guano in it. Procure real guano, and you will get real returns.

The fossil Infusoria (as they were formerly called) are now termed Diatomaceæ, and are found in various parts of the world; such as Bermuda earth, Berg-mehl from Norway, the deposit from Mourne Mountain in Ireland, &c. They are found in immense quantities, and afford the microscopist innumerable objects. The same treatment as that usually employed for the Diatomaceæ must be followed with these deposits; but as they are sometimes obtained in hard masses, disintegration is first necessary. To effect this, they are usually boiled for a short time in diluted *liquor potassæ*, which will soon cause the mass to fall into a mud-like deposit. Water must then be immediately added, in order that all further action of the *liquor potassæ* may be stopped, otherwise the objects searched for will be dissolved. For this reason it is necessary to understand what substance is being dealt with, because some deposits are much finer and are acted upon more readily than others.

In mounting these objects, some are so delicate that they are almost invisible when balsam is used with them; they are therefore usually mounted *dry*. Others, however, are much coarser, and may be mounted in balsam like the Diatomaceæ mentioned in Chapter IV.

The common Infusoria cannot be mounted dry with any great success, though a few may be placed upon the glass slide and allowed to dry naturally, when their characters will be very well shown. To obtain anything like a natural appearance, they must be put up in fluid as in Chapter V.

Next to the Diatomaceæ, no class of microscopic objects has been more looked into of late than the Foraminifera. These animals are almost all marine, having a jelly-like body enclosed in one or more chambers of shell, which is generally composed of carbonate of lime. The shells are made with minute orifices, through which the pseudopodia (false feet) are extended by which the animal is enabled to lay hold of anything to draw itself along. From the possession of these orifices they derive their name, as *foramen* means a door or opening. They have been found in every depth of sea hitherto sounded, each depth being abundant in certain species; the lowest beds containing the greatest number of specimens, though with less variation of kinds. In chalk they are found in a fossil state, and may readily be shown (see Chapter IV.); in limestone and other hard stones they are abundant, and some mountains are composed principally of these shells.

The methods of obtaining Foraminifera are various. Many may be found upon seaweeds, which should always be examined as soon as possible after gathering. They are found in masses upon some coasts where the waves have carried and left them; but they are to be found the most abundantly in sand or mud dredged from the bottom of the sea. They must, however, be cleaned and separated from the mass of impurity with which they are usually mixed. This may be done in various ways, according to the nature of the accompanying matter. If sand alone, as is frequently the case, the whole mass must be *thoroughly* dried, and then stirred up in clean water. The sand will soon subside by its own weight, but the chambers of the Foraminifera, being filled with air, will float upon the surface, and may be skimmed off. There is, however, one objection to this mode of proceeding—some of these objects are so minute, the chambers containing comparatively so small a quantity of air, that they sink and are cast away with the refuse sand. On this account it is preferable to take the trouble of searching certain soundings under the microscope, using the camel-hair pencil, or some other contrivance before mentioned, to extract those objects which are required.* To clean the Foraminifera, Professor Williamson advises the transfer of the specimens to an evaporating dish containing a weak solution of caustic potash. This must be boiled for some moments, when the organic matter will be entirely dissolved, and the calcareous shells left free from impurity. They must now be well washed in water, so that all alkaline matter may be entirely removed.

If the specimens are in *mud*, we must proceed in a different way:—Stir up the whole mass in water, and allow it to stand until the heavier portion has sunk to the bottom; the water may then be poured off and examined to see if there are any objects contained in it. This process must be repeated until the water come off quite clear, when (if the search is for Foraminifera only) the solution of caustic potash may be used as before mentioned. However the soundings, &c., are cleaned, it is necessary to assort them

* In searching any earth or sounding in order to take objects therefrom, no method presents the same facilities as the use of the finest camel-hair pencil, to which, after being drawn through the lips, any forms will adhere, and yet be readily detached upon the slide. After a little practice the smallest objects may be separated. Captain Lang, however, stated that he used a single hair or bristle dipped into gum and dried, after which a slight breath would restore its adhesive power. With a very fine hair pencil during one winter I mounted about 1,400 slides, each one picked out of sea soundings, many of which had from six to twelve specimens upon them. The readiness with which the objects adhered to the point and were detached when required, rendered the process much more pleasant than using a bristle with gum. As to the numbers of objects to be taken from any sounding, even imagination often fails. Plaucus, it is said, collected 6,000 shells of Foraminifera from an ounce of sand from the shore of the Adriatic. Soldani collected from less than an ounce and a half of rock from the hills of the Casciana, in Tuscany, 10,454 fossil shells. Several of these were so minute that 500 weighed only a grain. And D'Orbigny found 3,840,000 specimens in an ounce of sand from the shores of the Antilles.

under the microscope with the camel-hair pencil or other contrivance, as it is impossible to obtain them fit for mounting without undergoing this process.

The sea soundings taken by order of Government are drawn from the bottom in a kind of apparatus ingeniously made for the purpose, and the sand, mud, &c., are brought up in their original state. Common soundings, however, are taken by lowering a heavy piece of lead coated with tallow, which consequently brings up a small portion of the matter from the bottom. Mr. George Mosley, the late Secretary of the Manchester Microscopic Society, obtained numbers of the "scrapings" from the sounding leads. To make any use of these it is, of course, necessary to free them from all traces of the tallow. Mr. Dancer places the sounding in a basin and pours boiling water upon it, which causes the melted grease to rise to the surface. When cold, this may be removed, and the water carefully decanted. The operation may be repeated until no grease appears, when the water may be withdrawn and *liquor ammoniæ* used, which will form a soapy solution with any remaining grease. This must be treated with hot water for the final washing. Care must be taken lest the finer forms be carried away in decanting the washing liquid. Should it be wished to make certain as to this point, each washing should be examined under the microscope. In some cases the process of Mr. Dancer will prove sufficient. Mr. Dale, however, gives a method of accomplishing the same result, which is much more readily completed; and as no fault can be found with these results, I will here give it in full: It is now well known that one of the products obtained from the naphtha of coal-tar is a volatile, oily substance, termed *benzole* (or, by French chemists, *benzine*), the boiling-point of which, when pure, is about 180° Fahrenheit, and which is a perfect solvent of fatty substances. In a capsule, previously warmed on a sand-bath, Mr. Dale mixes with the tallow soundings some of this benzole, until diluted so as to run freely, pressing the lumps with a glass rod until thoroughly mingled; the solution and its contents are then poured into a paper filter, placed in a glass funnel; the capsule is again washed with benzole, until the whole of the gritty particles are removed into the filter. A washing-bottle is then supplied with benzole, and the contents of the filter washed to the bottom until the liquid passes off pure, which may be tested by placing a drop from the point of the funnel on a warm slip of glass or bright platinum, when, if pure, the benzole will evaporate without residue or tarnish; if grease be present, the washing must be continued until they are free from it. After rinsing through *weak* acid, or alcohol, for final purification, the calcareous forms will be ready for mounting.

The filter and its contents may be left to dry spontaneously, when the latter can be examined by the microscope. Should time be an object, rapid drying may be effected by any of the usual methods; one of which, recommended by Mr. Dale, is to blow a stream of hot air through a glass tube held in the flame of a Bunsen's burner. The lower the boiling-point of the benzole, the more readily can the specimens be freed from it. A commoner quality may be used, but it is more difficult to dry afterwards.

Pure benzole being costly, this may appear an expensive process; but, with the exception of a trifling loss by evaporation, the whole may be recovered by simple distillation. The mixture of tallow and benzole being placed in a retort in a hot-water, a steam, or a sand bath, the benzole will pass into the receiver, and the tallow or other impurities will remain in the retort. When the whole of the benzole has distilled over, which is ascertained by its ceasing to drop from the condenser, the heat is withdrawn and the retort allowed to cool before the addition of fresh material. Half a dozen to a dozen filters, each with its specimen, can be in process at the same time; and the distillation of the recovered benzole progresses as quickly as the filtration, which was practically proved on the occasion named. The process is very dangerous and great caution is to be used in the approach of light to the inflammable vapour.

After the Foraminifera and calcareous forms have been removed, the residue may be treated with acids and levigation in the usual manner, to obtain siliceous forms and discs, if there be any present; but to facilitate their deposition, and to avoid the loss of any minute atoms suspended in the washings, I would suggest the use of filtration. The conical filter is unsuitable, as the particles would spread over too great a surface of paper; but glass tubes open at both ends (such as broken test-tubes) will be found to answer, the broad end covered with filtering-paper and over that a slip of muslin tied on with a thread to facilitate the passage of the water and prevent the risk of breaking the paper. Suspend the tube over a suitable vessel through a hole cut in thin wood or cardboard, pour in the washings, which can be thus filtered and then dried. The cloth must be carefully removed, the paper cut round the edges of the tube, and the diatoms on the paper disc may be removed by a camel-hair pencil or otherwise, ready for mounting. Thus many objects may be preserved which would be either washed away or only be obtained by a more tedious process.

Such is Mr. Dale's method of cleaning the soundings from tallow, and as it thoroughly accomplishes its end, and is alike effective and not injurious to Foraminifera and diatoms, it may be safely recommended. The weak solution of caustic potash before advised for Foraminifera, must not be used where it is desired to preserve the diatoms, as they would certainly be injured, or destroyed altogether, if this agent were employed.

In fixing the Foraminifera upon the slide, no better plan can be followed than the dry cells and gum recommended in the early parts of this chapter. Owing to their thickness and composition, most of them are opaque objects only; but they are exquisitely beautiful, and require no particular care, except in allowing the cell to be perfectly dry when the cover is placed upon it, or the damp will certainly become condensed upon the inner side, and the examination seriously hindered.

Many of the Foraminifera require cutting into sections if it is wished to examine their internal structure;—"decalcifying" is also desirable in some cases: both of these processes will be found described at length in the chapter on Sections and Dissections.

When more than one specimen of some particular shell is obtained, it is better to place them upon the slide in different positions, so as to show as much of the structure as possible. I will conclude this subject by quoting a passage from T. Rymer Jones: "It is, therefore, by no means sufficient to treat these shells as ordinary objects by simply laying them on a glass slide, so as to see them only from one or two points of view; they must be carefully examined in every direction, for such is the diversity of form that nothing short of this will be at all satisfactory. For this purpose, they should be attached to the point of a fine needle, so that they may be turned in any direction, and examined by reflected light condensed upon them by means of a lens or side reflector. In many of the thick-shelled species it will be necessary to grind them down on a hone [see Chapter VI.] before the number and arrangement of the internal chambers is discernible; and in order to investigate satisfactorily the minutiæ of their structure, a variety of sections, made in various ways, is indispensable."

A visitor to the seaside may with little trouble procure one of the most beautiful objects which can ornament a cabinet. On turning over stones which have been covered by the last tide, a very small species of starfish will often be met with. From a small circular centre five long arms project, each of which is covered with spines beautifully arranged. When found, they should be dropped into fresh water with a little spirit added. This kills them instantly, else many of their long arms get broken by their struggles.

By putting them into water the arms are rendered, soft and may then be spread in forms best suited to microscopic slides, and thus allowed to dry. They are beautifully delicate in colour, needing no preparation to bleach them. During one morning's walk at Llandudno I procured about three dozen.

Plants afford an almost inexhaustible treasury for the microscope, and many of them show their beauties best when mounted dry. When any of these are to be mounted, care must be taken that they are thoroughly dry, otherwise the damp will certainly arise in the cell, and injure the object; and it may here be mentioned that long after a leaf has every appearance of dryness, the interior is still damp, and no way can be recommended of getting rid of this by any quicker process than that of keeping them in a warm room, as many leaves, &c., are utterly spoiled by using a hot iron or other contrivance. The safest way is to press them gently betwixt blotting-paper, which may be removed and dried at short intervals; and though this may appear a tedious operation, it is a *safe* one.

On the surface of leaves, hairs and scales of various and very beautiful forms are found, most of which display their beauties best when removed from the leaf, and used with the polarizer. These will be noticed in another place; but a portion of the leaf should always be prepared in its natural form, to show the arrangement of the hair or scales upon it; and this must almost invariably be mounted dry when used for this purpose. Many of them require very delicate handling. The *epidermis*, or, as it is by some termed, the *cuticle*, is the outer skin which lies upon the surface of the leaves and other parts of most plants. This is composed of cells closely connected, often bearing the appearance of a rude network. In many plants, by scraping up the surface of the leaf, a thin coating is detached, which may be torn off by taking hold of it with forceps. The piece may then be washed and floated upon a glass slide, where, on drying, it will be firmly fixed, and may usually be mounted dry. Amongst the most beautiful and easily prepared of these may be mentioned the petal of the geranium, the cells of which are well defined and amongst the most interesting.

Sometimes this *cuticle* is removed by maceration of the leaf in water or by a quicker method—boiling in nitric acid. Perhaps it will be as well to give Mr. Arnold's experience. "A leaf of a rhododendron which had been dry some months, and a freshly-gathered leaf of an azalea, were put into a test-tube, and covered with undiluted nitric acid of commerce of, I believe, about 1·32 specific gravity; the tube was held over a spirit-lamp until the acid just boiled, and the contents were then thrown into a basin of cold water. The cuticle of the rhododendron leaf partially separated spontaneously; that of the azalea came off without the least difficulty. The whole operation did not occupy more than five minutes. Undoubtedly many leaves, according to their texture, will require different strengths of acid, and longer or shorter periods of boiling."

Closely connected with the leaves are the ANTHERS and POLLEN, of which a great number are beautiful and interesting subjects for the microscopist.

The mallow tribe will furnish some exquisite objects, bearing the appearance of masses of costly jewels. These are usually dried with pressure, but the natural form may be more accurately preserved by allowing them to dry as they are taken from the flower, with no interference except thoroughly protecting them from all dust. Sometimes the anther is divided, so that the cell required to receive them may be of as little depth as possible. The common mallow is a beautiful object, but I think the lavatera is a better, as it shows the pollen-chambers well, when dried unpressed. The pollen is often set alone, and is well worth the trouble, as it then admits of more close examination. Often it is convenient to have the *anther and pollen* as seen in nature on one slide, and the *pollen* alone upon another. The former should be taken from the flowers before their full development is attained, as, if overgrown, they lose much of their beauty. Some pollens are naturally so dark that it is necessary to mount them in Canada balsam or fluid, as described in other places; but they are better mounted dry when they are not too opaque.

Here we may also mention the SEEDS of many plants as most interesting, and some of them very beautiful, objects, requiring for the greater part but a low power to show them. Most of these are to be mounted dry, as opaque objects, in cells suited to them, but some are best seen in balsam, and will be mentioned in Chapter IV.

The CORALLINES, many of which are found on almost every coast, afford some very valuable objects for the microscope. They must be well washed when first procured, to get rid of all the salts of the sea-water, dried and mounted in cells deep enough to protect them from all danger of pressure, as some of them are exceedingly fragile. The white ivory appearance which some of them present is given to them by an even covering of carbonate of lime; and should it be desired to examine the structure of these more closely, it may be accomplished by keeping them for some time in vinegar or dilute muriatic acid, which will remove the lime and allow of the substance being sliced in the same way as other Algæ. ("Micrographic Dictionary," p. 183.)

THE SCALES OF INSECTS.—The fine dust upon the wings of moths and butterflies, which is so readily removed when they are handled carelessly, is what is called the *scales*. To these the wing owes the magnificent colours which so often are seen upon it; every particle being what may be termed a distinct flat feather. How these are placed (somewhat like tiles upon a roof) may be easily seen in the wing of any butterfly, a few being removed to aid the investigation. Their form is usually that of the battledore with which the common game is played, but the handle or base of the scale is often short, and the broad part varies in proportionate length and breadth in different specimens. The markings upon these also vary, some being mostly composed of lines running from the base to the apex, others reminding us of network—bead-like spots only are seen in some—indeed, almost endless changes are found amongst them. These scales are not confined to butterflies and moths, nor indeed to the *wings* of insects. The different gnats supply some most beautiful specimens, not only from the wings, but also from the proboscis, &c.; whilst from still more minute insects, as the podura, scales are taken which are esteemed as a most delicate test. The gorgeous colours which the diamond beetles show when under the microscope are produced by light reflected from minute scales with which the insects are covered.

In mounting these objects for the microscope it is well to have the part of the insect from which the scales are usually taken as a separate slide, so that the natural arrangement of them may be seen. This is easily accomplished with the wings of butterflies, gnats, &c.; as they require no extraordinary care. In mounting the *scales* they may be placed upon slides, by passing the wings over the surface, or by gently scraping the wing upon the slide, when they must be covered with the thin glass. Of course, the extreme tenuity of these objects does away with the necessity of any cell excepting that formed by the gold-size or other cement used to attach the cover. The scales of the podura should be placed upon the slide in a somewhat different manner. This insect is without wings, and is no longer than the common flea. It is often found amongst the sawdust in wine-cellars, continually leaping about by the aid of its tail, which is bent underneath its body. Dr. Carpenter says:—"Poduræ may be obtained by sprinkling a little oatmeal on a piece of black paper near their haunts; and after leaving it there for a few hours, removing it carefully to a large glazed basin, so that, when they leap from the paper (as they will when brought to the light), they may fall into the basin, and may thus separate themselves from the meal. The best way of obtaining their scales, is to confine several of them together beneath a wine-glass inverted upon a piece of fine smooth paper; for the scales will become detached by their leaps against the glass, and will fall upon the paper." These scales are removed to the slide, and mounted as those from gnats, &c. When the podura has been caught without the aid of meal, it may be placed upon the slide, under a test-tube, or by any other mode of confinement, and thus save the trouble of transfer from the paper before mentioned. Another method is to seize the insect by the leg with the forceps and drag it across the slide, when a sufficient quantity of scales will probably be left upon it.

Mr. McIntyre procures the scales in the following manner:—He makes what he terms a breeding-cage, by taking a piece of plate-glass four inches long by two inches wide, and over this places a few sheets of blotting-paper. Upon these he lays a sheet of cork about a quarter of an inch thick, with a circle cut out of the centre one inch wide. This gives a kind of box, which he covers with glass, kept firm by two elastic bands. He says:—"After capturing the insect by means of a tube and a camel-hair pencil, I let it remain for some days in one of the breeding-cages, into which I always transfer the newly-caught podura, until it has changed its skin; then I stupefy it with chloroform, and drop it out on to a thin glass cover (previously cleaned) and with a very clean needle-point roll it backwards and forwards upon the cover till sufficient scales are removed. A *very light* pressure is indispensable, so as not to squeeze out any of the insect's fluids."

These scales are usually mounted dry; but Hogg recommends the use of Canada balsam (Chapter IV.) as rendering their structure more definite when illuminated with Wenham's parabolic reflector. Some advise other methods, which will be mentioned in Chapter V. As most *insects* when undissected are mounted in Canada balsam, the different modes of treatment which they require will be stated in another place.

In mounting blood of any kind to show the corpuscles, or, as they are often called, *globules*, which are round or oval discs, it is but necessary to cover the slide on the spot required with a coating as thin as possible and allow it to dry before covering with thin glass. There is a slight contraction in the globules when dried, but not enough to injure them for the microscope. The shape of these varies in different classes of animals, but the size varies much more, some being many times larger than others. Perhaps it will not be out of place to say a few words concerning the detection of blood. Wherever the stains are, they must be carefully scraped away and immersed for a few hours in a weak solution of bichloride of mercury. With a thin tube the more solid portion may then be removed to a glass slide and examined with a somewhat high power. A slight knowledge of the microscopic appearance of blood-discs will show us whether the suspicion of blood is correct.

Some of the skins of *larvæ* are beautiful objects; but, like many sections of animal and other fragile matter, are difficult to extend upon the slide. This difficulty is easily overcome by floating the thin object in clear water, immersing the slide, and when the object is evenly spread gently lifting it. Allow it then to dry by slightly raising one end of the slide to aid the drainage, and cover with thin glass as other objects. The tails and fins of many small fish may be mounted in a similar manner, and are well worth the trouble.

A few objects which are best shown by mounting *dry* may be here mentioned as a slight guide to the beginner, though some of them have been before noticed. Many of the Foraminifera, as elsewhere described. Some *crystals* are soluble in almost any fluid or balsam, and should be mounted *dry;* a few, however, deliquesce or effloresce, which renders them worthless as microscopic objects.

The wings of butterflies, gnats, and moths will afford many specimens wherewith to supply the cabinet of the young student. A great variety of scales also may be found amongst the ferns; indeed, these alone will afford the student occupation for a long time. On the under-side of the leaves are the reservoirs for the spores, which in many instances somewhat resemble green velvet, and are arranged in stripes, round masses, and other forms. The spores are usually covered with a thin skin, which is curiously marked in some specimens, often very like pollen-grains. The manner in which these spores with all their accompaniments are arranged, their changes and developments, afford almost endless subjects for study; different ferns presenting us with many variations in this respect totally invisible without the aid of the microscope. The hymenophyllums (of which two only belong to England) are particularly interesting, and the structure of the leaves when dried makes them beautiful objects, often requiring no balsam to aid their transparency. Portions of the *fronds* of ferns should be mounted as opaque objects, after having been dried between blotting-paper, when they are not injured by pressure; but care must be taken to gather them at the right time, as they do not show their beauty before they are ripe, and if over-ripe the arrangement of the spores, &c., is altered. The spores may be mounted as separate objects in the same manner as pollen, before mentioned, and are exquisitely beautiful when viewed with a tolerably high power. The number of foreign ferns now cultivated in this country has greatly widened the field for research in this direction; and it may also be mentioned that the under-sides of many are found to be covered with scales of very beautiful forms. A small piece of the frond of one of these may be mounted in its natural state, but the removal of the scales for examination by polarized light will be described in another place. The mosses also are quite a little world, requiring but a low power to show their beauties. The leaves are of various forms, some of which resemble beautiful net-work; the "urns" or reservoirs for the spores, however, are perhaps the most interesting parts of these objects, as also of the liverworts which are closely allied to the mosses. These urns are generally covered by lids, which fall off when the fruit is ripe. At this period they are well fitted for the microscope. The common screw-moss may be found in great abundance, and shows this denudation of the spores very perfectly. Many of these may be easily dried without much injury, but they should also be examined in their natural state.

The student should not omit from his cabinet a leaf of the nettle and the allied foreign species, the mystery of which the microscope will make plain. The hairs or stings may also be removed, and viewed with a higher power than when on the leaf, being so transparent as to require no balsam or other preservative.

There are few more interesting objects than the *raphides* or *plant-crystals*. These are far from being rare, but in some plants they are very minute, and consequently require care in the mounting, as well as a high magnifying power to render them visible; in others they are so large that about twenty-five of them placed point to point would reach one inch. Some of these crystals are long and comparatively very thin, which suggested the name (*raphis*, a needle); others are star-like, with long and slender rays; while others again are of a somewhat similar form, each ray being solid and short. If the stem of rhubarb, or almost any of the hyacinth tribe, be bruised, so that the juice may flow upon the slide, in all probability some of these crystals will be found in the fluid. To obtain them clean, they must be freed from all vegetable matter by maceration. After this they must be thoroughly washed and mounted dry. They are also good *polarizing* objects, giving brilliant colours; but when used for this purpose they must be mounted as described in Chapter IV. A few plants which contain them may be mentioned here. The Cactaceæ are very prolific; the orchids, geraniums, tulips, and the outer coating of the onion, furnish the more unusual forms.

The Fungi are generally looked upon as a very difficult class of objects to deal with, but amongst them some of the most available may be found. The forms of many are very beautiful, but are so minute as to require a high magnifying power to show them. The mould which forms on many substances is a fungus, and in some cases may be dried and preserved in its natural state. A friend of mine brought me a rose-bush completely covered with a white blight. This was found to be a fungus, which required a high magnifying power to show it. Being a very interesting object, it was desirable to preserve it, and this was perfectly effected without injury to the form by simply drying the leaf in a room usually occupied. Amongst the fungi are many objects well worth looking for, one of which is the *Diachœa elegans*. This, the only species, says the Micrographic Dictionary, is found in England upon the living leaves of the lily-of-the-valley, &c. These little plants grow in masses, reminding one of mould, to a height of a quarter of an inch, and each "stem" is covered with a sheath, in shape somewhat like an elongated thimble. When ripe the sheath falls off and reveals the same shaped column, made up of beautifully fine network, with the spores lying here and there. This dries well, and is a good object for the middle powers. Amongst the fungi the blights of wheat and of other articles of food may be included. Many of them may be mounted dry; others, however, cannot be well preserved except in liquids, and will be referred to in Chapter V. When rambling in a wood during the summer I sat down upon the fallen trunk of a tree, and here and there a few minute white spots caught my eye. I took my Coddington lens from my side-pocket and applied it to these. Judge of my surprise when I found each white speck a distinctly formed fungus resembling in size and form, to an amusing similarity, a disc of the arachnoidiscus. They were already dry, and I mounted them as ordinary dry objects; and hitherto no change has taken place which I can detect. Amongst the zoophytes and sea-mats, commonly called sea-weeds, may be found very many interesting objects to be mounted dry. When this mode of preservation is used, it is necessary that all the sea-salt be thoroughly washed from them. As they are, however, most frequently mounted in balsam or liquid, they will be more fully noticed in other places.

The *scales of fishes* are generally mounted dry when used as ordinary objects; but for polarized light, balsam or liquid must be used, as noticed in Chapter IV. To mount a fish-scale, however, in a satisfactory manner, care must be taken that it is perfectly clean. This can be accomplished only by careful washing, in which process soft camel-hair pencils will often be useful. When the slime or mucus has once dried, it is very difficult to remove. The variety and beauty of these are quite surprising to the novice. It is also very interesting to procure the skin of the fish when possible, and mount it on a separate slide to show how the scales are arranged. The sole is one of the most unusual forms, the projecting end of each scale being covered with spines, which radiate from a common centre, while those at the extremity are carried out somewhat resembling the rays of a star. One of the skates has a spine projecting from the centre of each scale, which is a very curious opaque object, especially when the skin is mounted in the manner described. The perch, roach, minnow, and others of the common fishes give the student good objects for his cabinet, and may be procured without difficulty. The scale of the turbot is a splendid object for the polariscope when mounted in balsam.

Insects which are very transparent, or have the "metallic lustre" with which any medium would interfere, are mounted dry. The diamond-beetle, before mentioned, is a splendid example of this; the back is generally used, but the legs, showing the curious feet, are very interesting objects. Indeed, amongst the legs and feet of insects there is a wide field of interest. When they are of a horny nature, it is best to dry them in any form preferred, but to use no pressure; when, however, they are wanted flat, so as to show the feet, &c., extended, they must be dried with a gentle pressure betwixt blotting-paper if possible. But this will be treated more fully in Chapter IV.

The *eyes of insects* are sometimes allowed to dry in their natural shape, and mounted as opaque objects; but generally they are used as transparencies in balsam or liquid, so the description of the treatment which they require will be deferred to Chapter IV.

Hairs, when not too dark, are sometimes transparent enough when mounted dry, but are usually mounted in balsam. These will be more fully noticed in another place, but there are some without which no cabinet is deemed in anywise complete. Many different species of bats, English and foreign, present us with hairs the form of which we should never have dared to imagine without microscopic aid. Other curious objects are found in the antennæ of crabs. You can also readily know whether you are being deceived when you buy what you deem a real sealskin or sable. From some of the common caterpillars I have obtained exquisitely beautiful slides, and a kangaroo is a true friend to an object-gatherer.

The hair of the ornithorhynchus is a very curious object, having a thin place in the middle of its length, and so presenting somewhat the appearance of a flail.

These are a few of the objects which are often mounted dry, but some of them should be shown in balsam or liquid also, and there is much difference of opinion as to the best way of preserving others. This, however, is explained by the fact, that the transparency which balsam gives, interferes with one property of the object, and yet develops another which would have remained invisible if preserved dry. The only method of overcoming this difficulty is to keep the object mounted in both ways, which is comparatively little trouble.

I may here mention that many prefer the lieberkuhn for the illumination of opaque objects; and a good background is gained by putting upon the under-side of the slide, immediately beneath the object, a spot of black varnish, which does not interfere materially with the light.

CHAPTER IV.

MOUNTING IN CANADA BALSAM.

The nature and use of this substance has been before spoken of, so that the method of working with it may be at once described.

Perfect dryness of the objects is, if possible, more necessary in this mode of mounting than any other, as dampness remaining in the object will assuredly cause a cloudiness to make its appearance in a short time after it is fixed. Where pressure does not injure the specimens, they are most successfully treated when first dried betwixt the leaves of a book, or in any other way which may prove most convenient, as noticed in Chapter III.

Before describing the methods of proceeding with particular objects, general rules may be given which should be observed in order to succeed in this branch of mounting.

As the object is to be thoroughly immersed in the balsam, it is evident that when it has once been covered, so it must remain, unless we again free it by a process hereafter mentioned, which is very troublesome; and on this account there must be nothing whatever in the balsam except the object. The inexperienced may think this an unnecessary caution; but the greatest difficulty he will meet with is to get rid of minute bubbles of air, perhaps invisible to the naked eye, which appear like small globules when under the microscope, and render the slide unsightly, or even worthless. Balsam dissolved in benzole will be found invaluable in mounting without air-bubbles; if a few are left in the specimen, by the next morning they will have entirely disappeared. In making this solution the balsam should be first boiled gently till on dropping a small quantity into water it is found to be as hard as resin, the softened and warm solution may be now poured into a bottle, and when cool the benzole added in sufficient quantity to make it of a desirable thickness. Ten objects out of eleven contain air, or at least are full of minute holes which are necessarily filled with it; so that if they should be immersed in any liquid of thick consistency, these cells of air would be imprisoned, and become *bubbles*. The air, then, must be removed, and this is usually accomplished by soaking for some time in turpentine, the period required differing according to the nature of the object. In some cases, the turpentine acts upon the colour, or even removes it altogether, so that it must be watched carefully. Often, however, this is an advantage, as where the structure alone is wanted, the removal of the colouring matter renders it more transparent. There are objects, however, which retain the air with such tenacity that soaking alone will not remove it. If these will bear heat without being injured, they *may* be boiled in turpentine, or even in balsam, when the air will be partly or totally expelled. But where heat is objectionable, they must be immersed in the turpentine, and so submitted to the action of the air-pump. Even with this aid, sometimes days are required to accomplish it perfectly, during which time the air should be exhausted at intervals of five or six hours, if convenient, and the objects turned over now and then.

Many complaints are made concerning turpentine, both as to its cleanliness and *penetrating* power. Most of these spring from the fact that few substances in the market vary so much as turpentine in purity; all sorts of rubbish are sold under this name, and now benzole is employed by many in all cases where turpentine alone was once used.

Sometimes the objects are so minute that it is impossible to submit them to any soaking, and in this case they must be laid upon the slide at once, and the turpentine applied to them there. But it must not be forgotten that there are some few which are much better mounted in such a way that the balsam may thoroughly surround, and yet not *penetrate*, the substance more than necessary. Sections of teeth are amongst these, but they will be noticed in another place, and some insects (see Dr. Carpenter) when required to show the ramifications of the tracheæ.

Having freed the object, then, from these two enemies—dampness and air—we now proceed to mount it.

The slide must first be cleaned; then on the centre a quantity of balsam must be placed with a bluntly-pointed glass rod, according to the size of the object about to be mounted. To this a slight heat must be applied, which will cause any bubbles to rise from the surface of the slide, so that they may be readily removed with a needle. The object having been freed from all air by steeping in turpentine, as before described, and then from superfluous liquid by a short drainage, or touch upon blotting-paper, is to be carefully laid *upon*, or where it is practicable thrust *into*, the balsam just prepared on the slide. In the former case, or where the balsam has not totally covered the object, a small quantity must be taken, warmed, and dropped upon it, and any bubbles removed by the needle as before. To cover this, the thin glass must be warmed, and beginning at one side, allowed to fall upon the balsam, driving a small "wave" before it, and thus expelling any bubbles which may remain. This is quite as safely performed (if not more so) by making a solution of balsam in turpentine of the consistency of thick varnish or by the use of chloroform and balsam, as mentioned in Chapter II. The thin glass cover may be slightly coated with this, and will then be much less liable to imprison any air, which frequently happens when the cover is dry. Bubbles, however, will sometimes make their appearance in spite of all care; but when the object is comparatively strong, they may be removed by keeping the slide rather warm, and *working* the cover a little, so as to press them to one side, when they should be immediately removed with a needle point, otherwise they are again drawn under.

Where the slide requires keeping warm for any length of

time, a *hot-water bath* is sometimes made use of, which is simply a flat tin, or other metal case, with a mouth at the side, that when the hot water is introduced it may be closed up, and so retain its warmth for a long time. An excellent bath may be made of an ordinary water-plate—costing about 1s. 9d. This may be filled either with hot water or sand, and if to it be added a flat tin cover such as is used in eating-houses, costing about 6d.—a very effective oven for baking slides is the result. It may be placed on the hob, or over or near any source of heat. It is easy to add a thermometer if necessary. In working, the slide is laid upon it, and so admits of longer operations, when required, without growing cold. Sometimes a spirit-lamp is placed under it to keep up an equal heat through excessively long processes. Where the time required, however, is but short, a thick brass plate is sometimes used (see Chapter II.), which may be heated to any degree that is required, the slide being previously placed upon it.

Some objects, which are so thin that they are usually *floated* upon the slide, as before stated, require no steeping in turpentine or other liquid. These are best mounted by covering with a little *diluted* balsam, and after this has had time to penetrate the substance, ordinary balsam is laid upon it, and the slide finished in the usual manner.

I have stated that balsam is usually applied to the slide and objects with a bluntly-pointed glass rod; but for the purpose of drawing the balsam from the bottle, and conveying it to the desired place, Dr. Carpenter uses a glass syringe with a *free* opening. These are his instructions:— "This (the syringe) is most readily filled with balsam, in the first instance, by drawing out the piston, and pouring in balsam previously rendered more liquid by gentle warmth; and nothing else is required to enable the operator at any time to expel precisely the amount of balsam he may require, than to warm the point of the syringe, if the balsam should have hardened in it, and to apply a very gentle heat to the syringe generally, if the piston should not then be readily pressed down. When a number of balsam objects are being mounted at one time, the advantage of this plan in regard to facility and cleanliness (no superfluous balsam being deposited on the slide) will make itself sensibly felt," but the collapsible metal capsules are certainly the best and most easily managed.

When the mounting has been thus far accomplished, the outer wall of balsam may be roughly removed after a few hours have elapsed; but great care is necessary lest the cover be moved or disturbed in any way. In this state it may be left for the final cleansing until the balsam becomes hard, which takes place sooner or later, according to the degree of warmth to which it has been subjected. Many advocate baking in a slow oven to accelerate this drying; but with some objects even this heat would be too great, and generally a mantel-piece, or other place about equal to it in temperature, is the best suited to this purpose; and when the requisite hardness is attained, the slide may be finished as follows:—With a pointed knife the balsam must be scraped away, taking care that the thin glass be not cracked by the point getting *under* it. If used carefully, the knife will render the slide almost clean; but any minute portions which still adhere to the glass must be rubbed with linen dipped in turpentine or spirit. If the balsam is not very hard, these small fragments are readily removed by folding a piece of paper tightly in a triangular form with many folds, and damping the point with which the glass is rubbed. As the paper becomes worn with the friction, the balsam will be carried off with it. In some cases I have found this simple expedient very useful.

Sometimes the object to be mounted is of such a thickness as to require a cell. For this purpose glass rings are used (as described in Chapter V.), and filled with balsam. The best mode of doing this is thus described by Mr. T. S. Ralph in the *Microscopic Journal*:—"The question was asked me when I was in England, if I knew how to fill a cell with Canada balsam and leave behind no air-bubbles? I replied in the negative; but now I can state how to accomplish this. Fill the cell with clear spirit of turpentine, place the specimen in it, have ready some balsam just fluid enough to flow out of the bottle when warmed by the hand; pour this on the object at one end, and gradually inclining the slide, allow the spirit of turpentine to flow out on the opposite side of the cell till it is full of balsam; then take up the cover, and carefully place upon it a small streak of Canada balsam from one end to the other. This, if laid on the cell with one edge first, and then gradually lowered until it lies flat, will drive all the air before it, and prevent any bubbles from being included in the cell. It can be easily put on so neatly as to require no cleaning when dry. If the cover be pressed down too rapidly, the balsam will flow over it, and require to be cleaned off when hardened, for it cannot be done safely while fluid at the edges."

Sometimes with every care bubbles are enclosed in the balsam, injuring objects which are perhaps rare and valuable. If the object will not be injured by heat, carefully warming the slide over a lamp will often set loose and remove these pests; but should heat be objectionable, or the bubbles too closely imprisoned, the whole slide must be immersed in turpentine until the cover is removed by the solution of the balsam; and the object must be cleansed by a similar steeping. It may then be remounted as if new in the manner before described.

The balsam and chloroform described in Chapter II. is thus used; and where the object is thin, the mounting is very easily accomplished. When the object is laid upon the slide with a piece of glass upon it, and the balsam and chloroform placed at the edge of the cover, the mixture will gradually flow into the space betwixt the glasses until the object is surrounded by it, and the unoccupied portion filled. The chloroform will evaporate so quickly that the outer edge will become hard in a very short time, when it may be cleaned in the ordinary way. Sometimes the balsam is dissolved in the chloroform without being first hardened; but this is only to render it more fluid, and so give the operator less chance of leaving bubbles in the finished slide, as, the thicker the medium is, the more difficult is it to get rid of these intruders.

It has been before noticed that some have objected to chloroform and balsam, believing that it became *clouded* after a certain time. Perhaps this may be accounted for in part by the fact that almost all objects have a certain amount of dampness in them. Others are kept in some preservative liquid until the time of mounting, and these liquids generally contain certain salts (Chapter V.). If this dampness, as well as all traces of these salts, however small, are not totally removed—the former by drying, the latter by repeated washings—the addition of chloroform will render the balsam much more liable to cloudiness than when balsam alone was used. It may safely be affirmed that benzole will be found in all cases a more valuable solvent for Canada balsam than chloroform.

This mode of employing the balsam, however, will not be always applicable, as *chloroform* acts upon some substances on which balsam *alone* does not. Some salts are even soluble in it, the crystals appearing after a few days or weeks, whereas in balsam alone they are quite permanent. Experience is the only guide in some cases, whilst in others a little forethought will be all that is required.

The particular methods used for certain objects may now be entered upon. Many of the Diatomaceæ and fossil Infusoria, as they are sometimes termed, are mounted dry, and cleaned in the way described in Chapter III. Others are almost always placed in balsam, except where they are intended to be used with the lieberkuhn and dark background, by which means some of them are beautifully shown. The usual way of mounting them in balsam is as follows:—Take a drop of the water containing them, place it upon the slide, and evaporate over the lamp, whilst with a needle they may be dispersed over any space desired. When they are thoroughly dry, drop a little balsam on one side, and exclude the bubbles. The slide may then be warmed to such a degree that the balsam, by lifting the glass at one end, will be carried over the specimens, which may then be covered with thin glass, made warm as before described. Where the objects are quite dry, and loose upon the glass, it requires great care in placing the cover upon them, otherwise they are forced to one edge, or altogether from under it, in the wave of the balsam. For this reason, Professor Williamson adds a few drops of gum-water to the last washing, which causes them to adhere sufficiently to the glass to prevent any such mishap.

Mr. T. G. Rylands's method differs in some degree from the above, and is, to use his own words, as follows:—Thick balsam is preferable, and the burnt covers (see Chapter III.) to be mounted are laid in a convenient position, with the diatoms upwards. The slides required having been carefully cleaned and marked on the under side with a ring of ink about half an inch in diameter by the aid of a turntable to point out the centre, a drop of benzole is applied by a large pin to the diatoms on the cover, so as to exclude the air from the valves and frustules. The slide is then held over the lamp, and when warm, a sufficiently large drop of balsam is put upon it, and heated until it begins to steam. If small bubbles appear, a puff of breath removes them, The slide being held slightly inclined from the operator, and the drop of balsam becoming convex at its lower edge, the cover is brought in contact with it at that point, gradually laid down, pressed with the forceps, and brought to its central position. When cool the superfluous balsam (if any) is removed with a heated knife-blade, the slide cleaned with a little turpentine, and finished by washing in a hand-basin with soap and water. In this process there is no delay if the balsam be sufficiently thick, as the slide may be cleaned off almost before it is cold.

It is now well known that from common chalk it is an

easy matter to obtain interesting specimens of Foraminifera. Scrape a small quantity of chalk from the mass and shake it in water; leave this a few minutes, pour the water away and add a fresh quantity, shake up as before, and repeat two or three times. Take a little of the residue, and spread it upon the slide, and when quite dry, add a little turpentine. When viewed with a power of two hundred and fifty diameters, this will generally show the organisms very well. If it is desired to preserve the slides, they may then be mounted in Canada balsam. Mr. Guyon, in "Recreative Science," observes that the accumulation of the powder, by the action of the rain or exposure to the atmospheric action, at the foot or any projection of the chalk cliffs, will afford us better specimens than that which is scraped, as the organisms are less broken in the former.

Take a piece of chalk, and with a *soft* tooth or nail brush, brush it under water, and then wash the sediment well till the water is not coloured, when the residue will be nearly all Foraminifera.

The above is the most simple method of obtaining Foraminifera from chalk, but is not so satisfactory when any number of perfect slides is wanted for the cabinet. I shall, therefore, give additional particulars from the experience of good men. But some specimens of chalk seem to have undergone such powerful action that no perfect forms are found in them. This accounts for that disappointment which I have now and then heard expressed. One student says,—Take about one ounce of chalk, place it in a quart bottle with about a pint of water, shake it, and after a few moments pour off the milky fluid to about one fourth. Add more fresh water, shake and pour off, waiting longer each time for it to settle. Continue ten or twelve times in one day, and repeat two or three times a day for a few days, and the result will be a sediment entirely composed of Foraminifera. If fragments of chalk remain, the bottle has not been sufficiently shaken.

Mr. Robertson uses a somewhat different method. Break

a lump of fresh chalk into pieces not larger than a walnut; then crush, but not *grind*, lest you destroy the forms, into a coarse powder that will pass a somewhat wide sieve. Tie this, as a pudding, in a stout piece of calico. Drop into water and allow the bundle to become saturated, and then knead with the hands. This will expel a quantity of milky water. From time to time, after allowing the fluid to drain off, the cloth should be untied, and retied more closely to the mass; and when the contents are reduced to about one-third of their original bulk, all large pieces of chalk, portions of spines of echini, &c., should be removed, lest they injure the more delicate forms. Care must be taken in the kneading when the greater portion of the chalk has escaped, and at last the bag only shaken until the water flows from it almost clear. The whole may then be transferred to a bottle of clear water and treated as before described. The results, Mr. Robertson says, will be satisfactory, and the chalk must be very poor in fossils if 2 lb. would not satisfy any microscopic observer.

When the Foraminifera are of a larger size, though transparent enough to be mounted in balsam, the air must be first expelled from the interior, otherwise the objects will be altogether unsatisfactory. To accomplish this they must be immersed in turpentine and submitted to the action of the air-pump. So difficult is it to get rid of this enemy, that it is often necessary to employ three or four exhaustions, leaving them for some time under each. When all air has given place to the turpentine, they must be mounted in the ordinary way.

Of all objects which are commonly met with, few are such general favourites as the POLYCYSTINÆ; and deservedly so. Their forms are most beautiful, and often peculiar—stars varying in design, others closely resembling crowns; the *Astromma Aristotelis* like a cross, and many whose shapes no words could describe. The greater part, perhaps, of those which are usually sold, are from the rocky parts of Bermuda; but they are also found in Sicily, some parts of

Africa and America. They are usually mounted in balsam. but are equally beautiful mounted dry, and used with the lieberkuhn. They require as much care in cleaning as the Diatomaceæ, but the process is a different one. Sometimes this is effected by simply washing until they are freed from all extraneous matter; but this is seldom as effectual as it should be. In the *Microscopic Journal* Mr. Furlong gives the following method of treatment as the best he knows:—

Procure—

A large glass vessel with 3 or 4 quarts of water.

New tin saucepan holding 1 pint.

2 thin precipitating glasses holding 10 oz. each.

Take 3 oz. of dry Barbadoes earth (lumps are best), and break into rather small fragments. Put 3 or 4 oz. of common washing soda into the tin and half fill it with water. Boil strongly, and having thrown in the earth, boil it for half an hour. Pour nine-tenths of this into the large glass vessel, and gently crush the remaining lumps with a soft bristle brush. Add soda and water as before, and boil again; then pour off the liquid into the large vessel, and repeat until nothing of value remains. Stir the large vessel with an ivory spatula, let it stand for three minutes, and pour gently off nine-tenths of the contents, when the shells will be left, partially freed only, like sand.

2ND PROCESS.—Put common washing soda and water into the tin as before, and having placed the shells therein, boil for an hour. Transfer to the large vessel as before, and after allowing it to stand for one minute, pour off. Each washing brings off a kind of "flock," which seems to be skins.

3RD PROCESS.—Put the shells in a precipitating glass and drain off the water until not more than ½ oz. remains. Add half a teaspoonful of bicarbonate of soda, dissolve, and then pour in gently 1 oz. of strong sulphuric acid. This liberates the "flock," &c., and leaves the shells beautifully transparent. Wash well now with water to get rid of all salts and other soluble matter.

Some of the large shells are destroyed by this method, but none that are fit for microscopic use. An oblique light shows these objects best.

These are sometimes treated in the manner described in Chapter III., where the diatoms are spoken of, but many forms are liable to be injured by this severe process.

It has been before stated that some of the zoophytes may be mounted dry, and others examined as opaque or transparent objects according to their substance. They are very interesting when examined in the trough whilst living, but to preserve many of them for future examination they must be mounted in some preservative medium. Sometimes this may be one of the liquids mentioned in Chapter V., but if possible they should be kept in balsam, as there is less danger of injury by accident to this kind of slide. This method of mounting presents some difficulties, but I think that all agree as to the trustworthiness of Dr. Golding Bird's information on the subject, which appeared in the *Microscopic Journal*. Of this, space forbids me to give more than a condensed account, but I hope to omit nothing of moment to the reader for whom these pages are written.

After stating that there are few who are not familiar with these exquisite forms, and have not regretted the great loss of beauty they sustain in drying, he informs us that from their so obstinately retaining air in the cells and tubes when dried, it is hardly practicable to get rid of it; and they also shrivel up very seriously in the process of drying. The following plan, however, he has found almost faultless in their preparation.

To preserve them with extended tentacles, they should be plunged in cold fresh water, which kills them so quickly that these are not often retracted.* The specimens should be preserved in spirit until there is leisure to prepare them;

* It has been stated that the best method of killing zoophytes is to drop alcohol, French brandy, or benzole into the salt water in which they are placed; as this will cause no retraction of tentacles if it be done gradually.

if, however, they have been *dried*, they should be soaked in cold water for a day or two before being submitted to the following processes:—

1. After selecting perfect specimens of suitable size, immerse them in water heated to about 120°, and place them under the receiver of an air-pump. Slowly exhaust the air, when bubbles will rise and the water appear to be in a state of active ebullition. After a few minutes re-admit the air and again exhaust, repeating the process three or four times. This will displace the air from most, if not all, of the class.

2. Remove the specimens and allow them to drain upon blotting-paper for a few seconds; then place them in an earthen vessel fitted with a cover, and previously heated to about 200°. This heat may be easily got by placing the vessel for a short time in boiling water, wiping it immediately before using, with a thick cloth. The specimens are then dropped into this, covered with the lid, and immediately placed under the receiver of the air-pump, and the air rapidly exhausted. By this means they are dried completely, and so quickly that the cells have no time to wrinkle.

3. In an hour or two remove them from the air-pump and drop them into a vessel of perfectly transparent camphine. This may be quite cold when the horny, tubular polypidoms, as those of the Sertulariæ, are used; but should be previously heated to 100° when the calcareous, cellular Polyzoa are the objects to be preserved. The vessel should be covered with a watch-glass and placed under the receiver, the air being exhausted and re-admitted two or three times.

4. The slide which is to receive the specimen should be well cleaned and warmed so as to allow the balsam to flow freely over it. This must be applied in good quantity, and air-bubbles removed with the needle-point. Take the polypidom from the camphine, drain it a little, and with the forceps immerse it fully in the balsam. The glass to be

laid upon it should be warmed and its surface covered with a thin layer of balsam, and then lowered gradually upon it, when no bubbles should be imprisoned. A narrow piece of card-board at each end of the object, for the cover to rest upon, prevents any danger of crushing the specimen.

This mode of mounting polypidoms, &c., seems to give almost the complete beauty of the fresh specimens. They are very beautiful objects when viewed with common light, but much more so when the polarizer is used (in the manner described a little farther on).

To the above instructions there can be little to add; but I may here mention that some young students may not be possessed of the air-pump, and on this account put aside all search for those specimens which need little looking for at the seaside. Many of these, however, though they lose some beauty by the ordinary mode of drying, will, by steeping for some time in turpentine, not only be freed from the air-bubbles, but suffer so little contraction that they are a worthy addition to the cabinet.

Another class of objects is the *spicula* met with in sponges, &c. These are often glass-like in appearance and of various shapes; many are found resembling needles (whence their name); some from the synapta are anchor-like, whilst others are star-like and of complex and almost indescribable combinations. As some of these are composed of silex and are consequently not injured by the use of nitric acid, the animal substance may be removed by boiling them in it. Those, however, which are calcareous must be treated with a strong solution of potash instead; but whichever way is used, of course they must afterwards be freed from every trace of residue by careful washing.

These spicules may be often found amongst the sand which generally accumulates at the bottom of the jars in which sponges are kept by those who deal in them, and must be picked out with a camel-hair pencil. The specimens obtained by this means will seldom if ever require any cleaning process, as they are quite free from animal matter.

In the former chapter were noticed those insects or parts of them which are usually mounted dry. When they are large and too opaque to admit of the dry treatment, they must be preserved in Canada balsam or fluid. The first of these may now be considered.

It may be here mentioned, that with these objects much heat must not be employed, as it would in some instances give rise to a cloudiness, and almost invariably injure them.

In killing the insect it is necessary not to rub or break any part of it. This may be performed by placing it in a small box half filled with fragments of fresh laurel-leaves, by immersion in turpentine or strong spirit, as also in solutions of various poisonous salts. After which it may be preserved for some time in turpentine or other preservative liquid (Chapter V.) until required. As an assistance to the student, I believe that I can do no better than give him the plan pursued by my friend Mr. Hepworth, whose specimens are in every way satisfactory; but when his method is used, the insects must not have been placed in turpentine for preservation :—

"After destroying the insects in chloroform or sulphuric ether (methylated being cheaper), wash them thoroughly in a wide-necked bottle, half-filled, with two or three waters; the delicate ones requiring great care. Then immerse them in liquid potash (or Brandish's solution, which is stronger than the usual preparation), and let them remain a longer or shorter time according to their texture. When ready to remove, put one by one into a small saucer of clear water, and with a camel-hair pencil in each hand press them flat to the bottom, holding the head and thorax with the left-hand brush, and applying pressure with the other from above, downwards, giving the brush a rolling motion, which generally expels the contents of the abdomen from the thorax. A minute roller of pith or cork might be used instead of the brush. In larger objects, use the end of the finger to flatten them. Large objects require more frequent washing, as it is desirable to remove the potash thoroughly, or crystals are apt to form after mounting. Having placed them on the slides with thin glass covers, tied down with thread,* dry and immerse them in rectified spirits of turpentine; place the vessel under the receiver of an air-pump, and keep it exhausted until the turpentine has taken the place of the air-bubbles; they are then ready for the application of the balsam. Larger objects may often with advantage be transferred to a clean slide, as during the drying there is considerable contraction, and an outline showing this often remains beyond the margin. When closely corked, they may remain in the spirits two or three months. As you take them from the bottle, wipe as much turpentine off as possible before removing the thread, and when untied carefully wipe again, placing the finger on one end of the cover whilst you wipe the other, and *vice versâ*. By this means you remove as much turpentine from under the cover as is necessary; then drop the balsam, thinned with chloroform (see Chapter II.), upon the slide, letting the fluid touch the cover, when it will be taken in between the surfaces by capillary attraction; and after pressing the cover down, it may be left to dry, or you may hold the slide over a spirit-lamp for a few seconds before pressing down the cover. If heat is not applied, they are much longer in drying, but are more transparent. If made too hot, the boiling disarranges the objects, and if carried too far, will leave only the resin of the balsam, rendering it so brittle that the cover is apt to fly off by a fall or any jar producing sufficient concussion. Never lift the cover up, if possible, during the operation, as there is danger of admitting air. A few bubbles may appear immediately after mounting, but they generally subside after a few hours, being only the chloroform or turpentine in a state of vapour, which becomes condensed."

* This applies to the more delicate, which will not bear transferring after being once spread out and dried.

This method of preparing and mounting insects I can strongly recommend as giving first-rate results; but where the specimens are small, they seldom need the soaking in caustic potash which larger ones must have. It is only necessary to leave them awhile in turpentine, especially when they have been first dried with gentle pressure between two glasses, and then mount with balsam in the ordinary way. With many, even of the larger insects, by soaking them in turpentine or oil of cloves for a longer time, they are made so much more transparent and even colourless, as to exhibit their internal organs (which are visible in layers, by the aid of the binocular microscope), the muscles of the legs, &c. They become also very beautiful objects for the polariscope.

Amongst the insect tribes there is abundant employment, especially for the lower powers of the microscope. But if the deeper wonders and beauties of the animal economy are to be sought out and studied, it is desirable that the various parts should be set separately, in order that they may receive a more undivided attention, as well as be rendered capable of being dealt with under the higher powers. We will therefore briefly consider the treatment which the different portions require.

The eyes of butterflies, and indeed of almost all insects, afford materials for a study which is complete in itself. When examined with a tolerably high power, instead of finding each eye with an unbroken spherical surface, it is seen that many are composed of thousands of hexagonal divisions, each being the outer surface of a separate portion termed the *ocellus*. In others these divisions are square; but in all there is a layer of dark pigment surrounding their lower parts. The ocelli may be partly removed from the eye, which will show how their tapering forms are arranged. But here we have to consider how to place them in balsam for preservation. The eye being removed from the insect, and the dark pigment removed by the use of a camel-hair pencil, must be allowed to remain in turpentine at least for some days. The turpentine should then be renewed and the eye well washed in it just before it is to be mounted. It may then be set in balsam in the same way as any other object;—but here a difficulty is met with. The eye being spherical upon the surface required, must necessarily be "folded" or broken in attempting to flatten it. This difficulty may be often overcome by cutting a number of slits round the edges; but some object to this mode of treatment, and, where it is practicable, it is much more satisfactory to mount one in the natural rounded form and another flat. Instead, however, of mounting the organ *whole*, four or five slides may be procured from each of the larger ones, such as those of the dragon-fly, &c.

The *antennæ* also are often mounted on separate slides, as being better suited for higher powers and more minute examination than when connected with the insect. These two projecting organs, issuing from the head, are jointed, and moveable at will. They differ very much in form amongst the various species, and are well worth the attention of the microscopist. They are usually mounted with the head attached, and perhaps they are more interesting when thus seen. Some few are very opaque; to prepare these the following method has been advised, though it is far better to view them as opaque objects:—

Bleach the antennæ by soaking in the following solution for a day or two:—

Hydrochloric acid, 10 drops.
Chlorate of potash, ½ drachm.
Water, 1 oz.

This will render them transparent. Wash well, dry, and mount in Canada balsam. Instead of the above, a weak solution of chloride of lime may be used, by which means the nerves will be well shown. Many, however, are rendered transparent enough by simply soaking in turpentine for a longer or shorter time. Where the antennæ, however, are "Plumose," or feather-like, extreme care is required in mounting, though the difficulty is not so great as some seem to think. If they are first dried with gentle pressure, and then subjected to the action of the air-pump in a small quantity of turpentine until the air is thoroughly expelled, they can be easily finished upon the slide, especially when balsam and chloroform are used.

Insects supply us with another series of beautiful objects, viz., the *feet.** These are sometimes simply dried and mounted without any medium, as before mentioned; but most of them are rendered much more fit for examination by using balsam in their preservation, as it greatly increases their transparency. The smaller kinds may be dried with gentle pressure betwixt blotting-paper, and then immersed for some days in turpentine, without requiring the treatment with liquor potassæ. This immersion will render them beautifully transparent, when they may be mounted in balsam, in the usual manner.

It is, however, sometimes found difficult to fix the feet when *expanded*, in which state the interest of the object is greatly increased. Mr. Ralph recommends the following mode:—"First wash the feet, while the insect is yet alive, with spirits of wine; then, holding it by a pair of forceps close to the edge of a clean piece of glass, the insect will lay hold of the upper surface by its foot; suddenly drop another small piece of glass over it, so as to retain the foot expanded, and cut it off with a pair of scissors, tie up and soak to get rid of air." Mr. Hepworth says that he never found any difficulty in expanding the foot *on* a drop of water or well-wetted slide, and laying a thin glass cover over it, tying with thread, drying, and immersing in turpentine.

The mouth, also, with its organs, is an interesting object in many insects. That of the common fly is often used, and is comparatively easy to prepare. By pressing the head, the tongue (as it is commonly termed) will be forced

* See Mr. Hepworth's interesting articles on the fly's foot in the second and third volumes of the *Microscopic Journal.*

to protrude, when it must be secured by the same means as the foot, and may be subjected to the soaking in turpentine, and mounted as usual. The honey-bee is, however, very different in formation, and is well worth another slide; indeed, even in insects of the same class, the differences are many and interesting. There is another good friend to the Microscopic Cabinet, the large water-beetle, "*Dyticus marginalis*"; and he is by no means uncommon, as he may be met with in many old ponds. If his wings are taken, dried, and mounted in balsam, beautiful circles with crosses make their appearance when examined by the aid of polarized light. But what are commonly termed his suckers are perhaps, his most popular gifts. On his anterior legs will be found small *discs* attached to central members (making the whole an exact resemblance of a boy's sucker), which may be readily cut off, placed on the slide, and mounted in balsam. The *Dyticus* also gives splendid examples of spiracles; but this will be mentioned where dissection is treated of.

Another worthy object of study is the *respiration* of insects, which is effected by tracheæ or hollow tubes, which generally run through the body in one or more large trunks, branching out on every side. These terminate at the surface in openings, which are termed *spiracles*, or breathing organs. The *tracheæ* often present the appearance of tubes, constructed of a spiral thread, somewhat resembling the spiral fibres of some plants. These are very beautiful objects, and are generally mounted in balsam, for which reason they are mentioned here; but as they evidently belong to the dissecting portion, they will be fully treated of in another place.

Amongst the parasitic insects a great variety of microscopic subjects will be found. As these are usually small, they may be killed by immersion in spirits of turpentine; and, if at all opaque, may be allowed to remain in the liquid until transparent enough, and then mounted in Canada balsam.

The acarida, or *mites* and *ticks*, are well known; none, perhaps, better than those which are so often found upon cheese. Flour, sugar, figs, and other eatables are much infested by them; whilst the diseases called the *itch* in man, and the *mange* in animals, are produced by creatures belonging to this tribe. These insects are sometimes mounted by simply steeping them in turpentine, and proceeding as with other insects. The Micrographic Dictionary gives the following directions as to mounting *parts* of these:—"The parts of the mouth and the legs, upon which the characters are usually founded, may be best made out by crushing the animals upon a slide with a thin glass cover, and washing away the exuding substance with water: sometimes hot solution of potash is requisite, with the subsequent addition of acetic acid, and further washing. When afterwards dried and immersed in Canada balsam, the various parts become beautifully distinct, and may be permanently preserved."

Feathers of different kinds of birds are usually mounted in balsam when required to show much of the structure. This is particularly interesting when the feathers are small, as they then show the inner substance, or *pith*, as it may be termed, with the cells, &c. The "pinnæ," or soft branches of the feathers, will be found of various constructions; some possessing hooks along one side, whereby they fasten themselves to their neighbours; others branching out, with straight points somewhat resembling the hairs from certain caterpillars. But, of course, when the metallic-looking gorgeous colours are all that is required to be shown, and reflected light used (as with the feathers of the humming-bird, peacock, &c.), it is much better that they should be mounted dry, as in Chapter III.

The *seeds* and *pollen* of plants are most frequently mounted dry, as mentioned in Chapter III.; but the more transparent of the former, and the darker kinds of the latter, are perhaps better seen in Canada balsam. There is nothing particular to be observed in the manipulation, except that the glass cover must be applied lightly, otherwise the grains may be crushed. There are some objects which cannot be shown in a perfect manner when mounted *dry*, but when immersed in balsam become so very transparent that they are almost useless. To avoid this, it has been recommended to stain the objects any colour that may be convenient, and afterwards mount in balsam in the ordinary manner.

Permanent dyes, however, for these minute objects are not so readily procurable. My friend Mr. Abbey showed me that what was permanent with vegetable matter of one kind was totally untrustworthy with another. The most useful that I have tried is Magenta, and the colour is a convenient one. Whatever is used for this purpose should be in solution, and the object steeped for awhile and afterwards thoroughly washed, in order that no superfluous salt may remain. There are many liquids now sold by every chemist which will help the student in this respect.

Most objects intended for the polariscope may be mounted in Canada balsam; but there are some exceptions to this. Many of the salts are soluble in this medium, or their forms so injured by it, that glycerine or oil has to be used (see Chapter V.): others must be left in the dry form, as before mentioned; and some few it is impossible to preserve unchanged for any length of time. *Crystals*, however, are amongst the most beautiful and interesting subjects for polarization; and it is very probable that, by the aid of the polariscope, new and valuable facts are yet to be made known. For one who finds pleasure in form and colour, there is a field here which will only open wider upon him as he advances; and instead of being in anywise a merely mechanical occupation, it requires deep and careful study. The little here said on the subject will show this in some degree.

With almost every salt the method of *crystallization* must be modified to obtain the best forms; I may even go further than this, and say that it is possible to change these forms to such a degree that the eye can perceive no relationship to exist between them. If a solution of sulphate of iron be made, a small quantity spread evenly upon a slide, and then suffered to dry whilst in a flat position, the crystals often resemble the fronds of the common fern in shape. But if, whilst the liquid is evaporating, it is kept in motion by stirring with a thin glass rod, the crystals form separately, each rhombic prism having its angles well defined, and giving beautiful colours with the polarized light. Again, pyro-gallic acid, when allowed to flow evenly over the slide in a saturated solution, covers the surface in long needles, which are richly coloured by polarized light; but if any small portion of dust or other matter should form a nucleus around which these needles may gather, the beauty is wonderfully increased. A form very closely resembling the "eye" of the peacock's tail, both in form and colour, is then produced, which to one uninitiated in crystallography bears very little resemblance to the original crystal. From these simple facts it will be clearly seen that in this, as in every other department, study and experience are needful to give the best results.

By dropping a saturated solution of any salt into alcohol —where it is not soluble in the alcohol—crystals are instantaneously produced, and the results are often very curious and beautiful. These crystals can easily be taken up by a pipette—deposited upon a slide, and, after having been allowed to dry spontaneously—mounted in balsam.

To obtain anything like uniformity in the formation of crystals upon the glass slide, every trace of grease must be removed by cleaning with liquor potassæ or ammoniæ immediately before using, care also being taken that none of the agent is left upon the slide, otherwise it may interrupt and change their relative position, and even their form.

Amongst those which are generally esteemed the most beautiful, are the crystals of oxalurate of ammonia. The preparation of this salt from uric acid and ammonia is a rather difficult process, and will not, on that account, be described here; but when possessed, a small quantity of a strong solution in water must be made, and a little placed on the slide, and evaporated slowly. Part of the salt will then be deposited in circles with the needle-like crystals extending from common centres. They should then be mounted in pure Canada balsam; and, when the best colours are wanted, used with the selenite plate. Of this class of crystal, salicine is a universal favourite, and can be easily procured of most chemists. The crystals may be produced in two ways:—A small portion of the salt must be placed upon the slide, and a strong heat applied underneath until fusion ensues; the matter should then be evenly and thinly spread over the surface. In a short time the crystals will form, and are generally larger than those procured by the following process; but the uncertainty is increased a little when fusion is used, which, however, is desirable with many salts. Secondly, make a saturated solution of salicine, which is effected by boiling one part of the salt in eighteen of water, and allowing it to cool. Place a little upon the slide, and let it evaporate spontaneously, or with the aid of gentle heat. The crystals are generally uniform, and with ordinary powers quite large enough to make a beautiful object. Their circular shape and gorgeous colours—even without a selenite plate—have made them such great favourites that there are few cabinets without them.

There are also some salts which are crystallized in a somewhat different manner from those before mentioned. *Santonine* is one of the most beautiful, and will illustrate my meaning. Place a small portion upon a slide, and heat over a lamp until it is fused. With a *hot* needle spread the salt over the surface required. As the slide cools, the formation of crystals takes place, until it becomes one mass. This salt is slightly soluble in the ordinary balsam, and should be mounted in castor oil. If, however, the slide is well covered under the thin glass, the balsam soon becomes saturated, and very little injury results. According to the temperature during crystallization the character of the crystals is affected. If the fused salt is *very hot*, the crystals run in straight lines from a common centre. If the heat is (what I may term) *medium*, the crystals show concentric waves of very decided form. If the slide is *cool*, the crystals, still concentric, are exceedingly minute. The most beautiful crystals for the microscopist are those formed at a temperature betwixt the second and third above mentioned, as the minute and wavy forms are then combined, and long feathery crystals are the result. As this method requires some little practice, many crystallize the salt in a simpler manner, which I will give; but the variations obtainable by fusion give that mode the precedence. Dissolve a few grains of santonine in a drachm of chloroform, and drop the solution upon a glass slide. Allow the liquid to evaporate, and beautiful crystals will be the result. Mount as above.

In fear of being somewhat uninteresting to part of my readers, I feel as though I should not be fulfilling my desire of giving every information, if I omitted to show another method of crystallization, which a novice would cast away as a failure before he had completed his experiment. *Tartrate of soda*, made by neutralizing a strong solution of tartaric acid by the addition of carbonate of soda, is spread in solution over a glass slide, and must be then warmed, but not boiled. It must now be laid in a dry place, protected from all chance of dust. In time—from one or two days to as many weeks—some of the slides will prove beautiful objects, showing the cross form surrounded by rays. Some of these slides never crystallize, though I can find no reason for this, and even the application of heat to these calls out no decided form.

Hippuric acid will be found most interesting to those who are fond of beautiful polariscopic effects, inasmuch as this salt is capable of giving an astounding variety in the forms of its crystals. Make a saturated solution in absolute alcohol, and use it warm; by dropping a small quantity from a warm pipette on to a warm slide a film will spread over the slide and crystals of a circular form will begin to appear, and may be modified by the atmosphere in which they are allowed to grow; thus a moist atmosphere or the reverse, an atmosphere of vapour of ammonia, spirit, benzole, or sulphureous fumes, will each produce a different result, and the modifications thus produced will afford food for very serious reflection on the changes one salt may be made to assume in contact with other agents. These crystals are best mounted in castor oil—balsam that is very liquid—*not balsam in benzole,* as the benzole changes the character of the crystal.

Many new forms may be procured by uniting two totally different salts in solution in certain proportions. This is a field affording new facts and beauties; but requires some chemical knowledge and much perseverance to obtain very valuable results. One of the most beautiful I have met with has been composed of sulphate of copper and sulphate of magnesia. The flower-like forms and uniformity of crystallization when successful make it well worth a few failures at first; and as I became acquainted with some new facts in my frequent trials, I will give the preparation of the double salt from the beginning.

Make a saturated solution of the two sulphates, combined in the proportion of three parts copper to one part magnesia, and then add to the solution one-tenth of pure water. No dust or other impurities should have access to the slide, and it should be freed from all traces of grease by cleaning immediately before use with liquor potassæ or ammoniæ. A drop of the solution should then be placed upon it, and by a thin glass rod spread evenly upon the surface. Heat this whilst in a horizontal position until the salt remains as a viscous transparent substance, which will not be effected until it is raised to a high degree. The slide may now be allowed to cool, and when this is accomplished, the flower-like crystals will be perceived forming here and there upon the plate. When these are at any stage in which it is wished to preserve them, a few seconds' exposure to the fire, as warm as the hand can comfortably bear, will stop the expansion, when the portion which we wish to mount should be cut off from the mass of salt by simply scratching the film around, and pure Canada balsam with the thin glass used. Breathing upon the film, or allowing the slide to become cold and attract the moisture from the atmosphere, will cause the crystallization to extend, and sometimes greatly rob the effect; so it is necessary to mount quickly when the desired forms are obtained. As the crystals are very uncertain as to the place of their formation, I may here mention that they may be got in *any* part of the slide by piercing the film with a needle-point; but in some degree this necessarily interferes with the centre. Into the cause of this we have no need to enter here, and as it has been elsewhere discussed, I can only give the above directions, and say that there is a great field in this branch of study which the microscope alone has opened.

It would be useless to enter into particulars respecting the various salts and treatment they require, as a great difference is effected even by the strength of the solution. There are some crystals, also, which are called forth in insulated portions, showing no formation upon the ground; but even when mounted in any preserving fluid, and unchanged for a year, a new action seems to arise, and a groundwork is produced which bears little resemblance to the original crystal. Sometimes this new formation adds to the beauty of the slide; in other cases the reverse is the result, the slide being rendered almost worthless. This action, I believe, frequently arises from some liquid being contained in the balsam or other mounting medium used; and this is rendered the more probable by the crystallization being called forth in an hour after the balsam diluted with chloroform is employed, whereas no change would have taken place for months (if at all) had pure balsam been used.

Sections of some of the salts are very interesting objects; but the method of procuring these and their nature will be described in Chapter VI.

Few objects are more beautiful with polarized light than young *oysters*. Good colours and a decided cross are given by them when well prepared. The following is the method pursued by Mr. Henry Lee:—Having found a "black-sick" oyster (to use the dredgemen's term), the spawn of which is quite mature and ready for extrusion, pour off from the shell the dark slate-coloured fluid into a long narrow two-ounce phial; fill up the bottle with *distilled* water; shake it gently; allow the deposit to settle, and change the water two or three times, repeating the agitation to get rid of the salt. Then substitute for the water liquor potassæ, diluted with equal quantity of distilled water. Allow the young oysters to remain in this for two days, agitating occasionally; and as often as the liquid becomes discoloured pour it off, and renew the same until no colour is given off and the shells are seen to be thoroughly cleansed from all animal matter by their sinking freely and rapidly to the bottom. When this stage is arrived at, stop the process, that the two valves of the shells may not be separated by the destruction of the hinge. Wash repeatedly in distilled water, to remove all trace of alkali, and finally wash and preserve the shells in a little rectified spirits of wine (not methylated spirit). These objects are frequently mounted in balsam, to increase their transparency for the polariscope, but where they are sufficiently clear they may be mounted dry like the foraminifera.

The scales of various fish have been before mentioned as mounted dry; when, however, they are required for polarizing objects, they are generally mounted in balsam, and some few in liquid. The former method will be considered here.

The eel affords a beautiful object for this purpose. The scales are covered by a thin *skin*, which may be slightly raised with a knife and then torn off, in the same manner as the covering of the geranium and other petals, described in Chapter III. The required portion may then be removed; or if a piece of skin can be procured as stripped off in cooking, the scales may be easily taken from the inner surface. They must then be washed and thoroughly cleaned. After drying, soak for a day in turpentine, and mount in the ordinary manner with balsam. This is a good polarizing object; but the interest, and I think the beauty, is increased by procuring a piece of eel's skin with the scales *in situ*, washing and drying under pressure, and mounting in balsam as before. The arrangement of the scales produces beautiful "waves" of colour, which are quite soothing to the eye after examining some of the very gorgeous salts.

There are many scales of fish which are good subjects for the polariscope when mounted in balsam; but as they require no particular treatment, they need no mention by name.

Among hairs we find some which are beautiful when mounted in balsam and examined by polarized light. Some, when wanted as common objects, are always used dry, as before mentioned; but if they are intended to be shown as *polarizing* objects, they must be placed in some medium. The Micrographic Dictionary mentions a mode of making an interesting object by plaiting two series of white horsehairs at an angle, mounting in balsam, and using with the polariscope. All hairs, however, must be steeped in turpentine for a short time before mounting, as they will thus be rendered cleaner and more transparent. When this is done, there is no difficulty in mounting them.

Many of the "tongues" of fresh-water and marine mollusca are deeply interesting and most beautiful objects when examined by polarized light. As these are usually mounted in balsam, I mention them in this place; but as they must be removed from the animals by dissection, particulars respecting them will not be entered into until we come to the part in which that operation is described (Chapter VI.).

The manner of preparing and mounting many of the Polyzoa and Zoophytes has been before described; but any notice of *polarizing* objects would be incomplete without some allusion to them. A small piece of the *Flustra avicularis*, well prepared, is beautiful when examined in this manner. No selenite is needed, and yet the colours are truly gorgeous. It is often met with upon shells and zoophytes of a large size, and will well repay the trouble of searching for. Many of the Sertularidæ are very beautiful with polarized light, and, indeed, no ramble upon the seaside need be fruitless in this direction.

The different *starches* are quite a study in themselves, and are peculiarly connected with polarized light. They are found in the cellular tissue of almost every plant in small white grains, which vary considerably in size; that from the potato averages one-three-hundredth of an inch in diameter, and that from arrow-root about one-six-hundredth. To procure starch from any plant, the texture must first be broken up or ground coarsely; the mass of matter must be then well washed in gently-flowing water, and, as all starch is totally insoluble in cold water, the grains are carried off by the current and deposited where this is stayed. In procuring it from the potato, as well as many other vegetables, it is but necessary to reduce the substance to a coarse pulp by the aid of a culinary grater; the pulp should then be well agitated in water, and allowed to rest a short time, when the starch will be found at the bottom, its lighter colour rendering it easily distinguishable from the pulp. It should, however, be washed through two or three waters to render it perfectly clean.

These grains have no crystalline structure, but present a very peculiar appearance when examined with polarized light. Each grain shows a dark cross whose lines meet at the point where it was attached to the plant, called the *hilum*. Round the grain also a series of lines is seen, as though it were put together in plates. This is more distinctly visible in some kinds than others.

As to the mounting of these starches there is little to be said. If the grains are laid upon the slide, and as small a portion as possible of the balsam diluted with turpentine, as before mentioned, be applied, they will cling to the glass and allow the pure balsam to flow readily over them without being so liable to imprison air-bubbles when the thin glass is put upon them.

The raphides, which were fully described in Chapter III., when required for use with polarized light, must be mounted in balsam, and many are found which give beautiful colours. They require no peculiar treatment, but must be washed quite clean before putting up. But in order to understand anything of the natural arrangement of *raphides*, it is necessary to mount certain parts of plants with these objects *in situ*. The most common is the coating of the onion, which must be soaked some time in turpentine or benzole, in order to render it transparent, and must then be mounted in balsam, as before said. We shall then be able to obtain such colours by the aid of polarized light, that the *raphides* are shown in wonderful distinctness, and somewhat of their nature will be perceived.

There is one class of objects for the polariscope which differs in preparation from any we have yet considered, and affords very beautiful specimens. Some of the plants, including many of the grasses and the Equisetaceæ (*i.e.* horsetails), contain so large a quantity of silica, that when the vegetable and other perishable parts are removed, a skeleton of wonderful perfection remains. This skeleton must be mounted in balsam, the method of performing which will now be considered.

Sometimes the cuticle of the equisetum is removed from the plant, others dry the stem under pressure, whilst the grasses, of course, require no such preparation. They should then be immersed in strong nitric acid and boiled for a short time; an effervescence will go on as the organic matter is decomposed, and when this has ceased, more acid should be added. At this point the modes of treatment differ; some remove the object from the acid and wash, and having dried, burn it upon thin glass until all appears *white*, when it must be carefully mounted in balsam. I think, however, it is better to leave it in strong acid until all the substance, except the required portion, is removed; but this will take a length of time, varying according to the mass of the plant. Of course, when this latter method is used, the skeleton must be washed from the acid, &c., before being mounted in balsam.

These *siliceous cuticles* are readily found. The *straws* of most of the cereals, wheat, oat, &c.; the *husks*, also, of some of these; many *canes;* the equisetum, as before described; and some of the grasses. Many of these are everywhere procurable, so that the student can never want material for a splendid object for the polariscope.

In Chapter III. the *scales* (or hairs) which are often found upon the leaves of plants were mentioned as beautiful objects when mounted dry; but some of these when detached from the leaf—which is easily done by gently scraping it, when dried, with a knife—present brilliant starlike and other forms, if mounted in balsam and used with the polariscope. There is a little danger, when placing the thin glass upon the balsam, of forcing out the scales in the wave of matter which is always ejected; this may be overcome by applying to the slide, previously to placing the objects upon it, an extremely thin covering of balsam diluted with turpentine as before mentioned, letting it dry more or less with the objects placed in it, and then, after the addition of a little more balsam, putting the cover on, and thus giving them every chance of adherence; or by using the balsam with chloroform, as before noticed. This method is peculiarly successful in cases where it is desired to arrange several objects symmetrically on a slide, and to obviate their subsequent disturbance by placing the cover on. Type slides with several parts of an insect displayed upon them, scales of fish, or of plants, &c., may thus be shown, so that the *number* of slides may by this plan be seriously diminished.

These scales are much more abundant than was formerly supposed, and new specimens are discovered daily; so that the student should always be on the look-out for them in his researches in the vegetable world.

Most classes of objects, and the treatment they require when mounting them in balsam, have now been considered. The next chapter will be devoted to preservative liquids, and the best method of using them.

CHAPTER V.

PRESERVATIVE LIQUIDS, ETC., PARTICULARLY WHERE CELLS ARE USED.

There are many objects which would lose all their distinctive peculiarities if allowed to become dry, especially those belonging to the fresh-water Algæ, many animal tissues, and most of the very delicate animal and vegetable substances in which structure is to be shown. These must be preserved by immersion in some fluid; but it is evident that the fluid must be suited to the kind of matter which it is intended to preserve. As it often requires much study and trouble to *obtain* microscopic objects of this class, it is well that their *preservation* should be rendered as perfect as possible; and for this reason the cells, or receptacles of the fluids, should be so closed that all possibility of escape should be prevented. The accomplishment of this is not so easy a matter as it might appear to the inexperienced.

Before giving any directions as to the manipulation required in mounting the objects, we must consider the different *liquids* and *cells* which are requisite for their preservation. Of the former there are a great number, of which the principal may be mentioned.

Distilled Water is strongly recommended by many for Diatomaceæ and other Protophytes. It has been, however, stated that confervoid growths often disturb the clearness of the liquid, and on this account various additions are made to it. A lump of camphor is often left in the bottle, so that the water may dissolve as much as possible. One grain of bay salt and one of alum are added to each ounce of water; or a drop or two of creosote shaken up with an ounce of water, which should be afterwards filtered. These additions are often made; perhaps each of them good for certain objects.

GLYCERINE.—Some affirm this to be one of the best preservative liquids, especially for vegetable objects; but others think that it is much better when diluted with two parts of camphor-water, prepared as above.* Mr. A. E. Verrill, of Yale College, U.S., says glycerine preserves the natural colours of marine animals; and the only precaution to be taken is to use *very heavy* glycerine, and to keep up the strength by transferring the specimens to new as soon as they have given out water enough to weaken it much, repeating the transfer till all the water is removed before finally mounting on the slide.

GLYCERINE AND GUM.—This is also believed to be a very good liquid for vegetable tissues, and is thus prepared :—

Pure gum-arabic...	1	oz.
Glycerine	1	„
Water (distilled)...	1	„
Arsenious acid ...	1½	grain.

Dissolve the arsenious acid in the cold water, then the gum, add the glycerine, and mix without bubbles.

Dr. Carpenter states that the proportions used ultimately by the late Mr. Farrants are :—

Picked gum-arabic ...	4	parts by	weight.
Distilled water (cold)..	4	„	„
Glycerine...	2	„	„

Thus he now omits the arsenious acid, but places in the solution (which should be kept in a bottle with glass stop-

* Dr. Carpenter says :—"Glycerine has a solvent power for carbonate of lime, and should not be employed when the object contains any calcareous structure. In ignorance of this fact, the author (Dr. C.) employed glycerine to preserve a number of remarkably fine specimens of the pentacrinoid larva of the Comatula, whose colours he was anxious to retain ; and was extremely vexed to find, when about to mount them, that their calcareous skeletons had so entirely disappeared, that the specimens were completely ruined."

per) a small piece of camphor. This requires no cell, as the adhesive power is sufficient.

DEANE'S COMPOUND.—This is usually deemed about the best medium for preserving Algæ, mosses, &c., and is thus prepared:—Soak 1 oz. of best gelatine in 4 oz. of water until the gelatine becomes soft, when 5 oz. of honey heated to boiling point are added; boil the mixture, and when it has cooled, but not enough to become stiff, add ½ oz. rectified spirit with which 5 or 6 drops of creosote have been well mixed, and filter the whole through fine flannel. This compound when cold forms a stiff jelly, the use of which will be described elsewhere.

GLYCERINE JELLY.—This mixture closely resembles the above, but as the composition differs a little it may be mentioned here. It is strongly recommended by Mr. Lawrance in the *Microscopic Journal*, where he states "that the beautiful green of some mosses mounted two years ago, is still as fresh as on the day they were gathered;" and that this is the only medium he knows which will preserve the natural colour of vegetable substances. He takes a quantity of Nelson's gelatine, soaks it for two or three hours in cold water, pours off the superfluous water, and heats the soaked gelatine until melted. To each fluid ounce of the gelatine, *whilst it is fluid but cool*, he adds a fluid drachm of the white of an egg. He then boils this until the albumen coagulates and the gelatine is quite clear, when it is to be filtered through fine flannel, and to each ounce of the clarified solution add 6 drachms of a mixture composed one part of glycerine to two parts of camphor-water.

At the Academy of Natural Sciences of Philadelphia, Mr. W. H. Walmsley stated, that, owing to the heat of that climate, the above formula for glycerine jelly was not satisfactory, and recommended the following:—Take one package of Cox's gelatine, wash repeatedly in cold water; allow it to soak in water sufficient to cover it for an hour or two, add one pint of boiling water, and boil ten or fifteen minutes. Remove, and when cool but still fluid, add the white of an egg, well beaten, and again boil, until the albumen coagulates. Strain whilst hot through flannel, and add an equal portion by measurement of Bowyer's pure glycerine, and fifty drops of carbolic acid in solution; boil again for ten or fifteen minutes, and again strain through flannel, place in a water-bath and evaporate to about one-half, then filter into two or more broad-mouthed vials. (Cotton is the best filtering medium.) The use of this in mounting is the same as Mr. Lawrance's, elsewhere described.

GOADBY'S FLUID.—This is much used in the preservation of animal objects; and seldom, if ever, acts upon the colours. It is thus prepared:—Bay salt, 4 oz.; alum, 2 oz.; corrosive sublimate, 4 grains. Dissolve these in two quarts of boiling water, and filter. For delicate preparations some recommend that this mixture be reduced by the addition of an equal quantity of water; but where there is bone or shell in the object, the above acts injuriously upon it, in which case this fluid may be used:—Bay salt, 8 oz.; corrosive sublimate, 2 grains; water, 1 quart.

THWAITES'S LIQUID.—This is recommended for the preservation of Algæ, &c., as having little or no action on the colour, and is thus prepared:—Take one part of rectified spirit, add drops of creosote enough to saturate it; to this add sixteen parts of distilled water and a little prepared chalk, and filter. When filtered, mix with an equal quantity of camphor-water (as before mentioned), and strain through fine muslin before using.

CHLORIDE OF ZINC SOLUTION.—In the Micrographic Dictionary this is stated to be "perhaps the best preservative known for animal tissues." Persons of great experience, however, have given a very different opinion; but it is certainly very useful in many cases where a small degree of coagulating action is not injurious. It is used of strengths varying according to the softness of the parts to be preserved; the average being 20 grains of the fused chloride to 1 oz. of distilled water. To keep this liquid, a lump of camphor may be left floating in the bottle. I have heard complaints that this mixture becomes turbid with keeping; but I think this must only be the case when some impurity has got into the bottle.

Carbolic Acid.*—The addition of a few drops of this to distilled water prevents the growth of interfering substances which would take place if pure water alone were used, and is therefore valuable as a preservative fluid. The same solution also is convenient, as it instantly kills infusoria, and almost everything that has life; and, indeed, is useful in the student's gathering-bottles for the same reason. It is very highly spoken of as forming one of the constituents in the following formula for use in mounting soft animal textures:—

Arsenious acid, 20 parts.
Crystallized carbolic acid, 10 parts.
Alcohol, 300 parts.
Distilled water, 700 parts.

The Rev. W. W. Spicer, in his translation of Johann Nave's work on Algæ, recommends the following fluid for their preservation:—Pure alcohol, 3 parts; distilled water, 2 parts; glycerine, 1 part. If the desmid or other alga be placed in this fluid in a cell, and not covered by a glass for a time, the water and alcohol will evaporate slowly, and the mixture will become more dense in proportion, but quite gradually, and therefore without any destructive influence on the object. During this operation, water is withdrawn from the frustule, and the glycerine, which is not volatile, takes its place without causing any distortion of the object.

Castor Oil.—This is a very useful preservative for crystals and other objects. Many salts are quite destroyed when Canada balsam is used with them; but very few are acted upon by this oil. To use it, it must be dropped in sufficient quantity to cover the crystal or object to be

* Dr. Crace Calvert stated at a meeting of the British Association, that after careful experiments he finds carbolic acid "prevents the development of protoplasmic and fungoid life."

preserved with a thin coating of oil. It may be necessary sometimes to spread it with a needle or other instrument. The thin glass should then be carefully placed upon it, so that all air may be excluded; and should any oil be forced out, owing to the quantity used being too great, it must be removed with blotting-paper. When the edge of the thin glass cover and the surrounding parts of the slide are as clean as possible, a coating of sealing-wax varnish or liquid glue must be applied and allowed to dry. A second or even a third coating may be required, but not before the previous cover is quite dry. These varnishes, however, are very brittle, and it is much safer, as a finish, to use one of the tougher cements—gold-size, for instance—which will render it doubly secure.

The above are the principal liquids, &c., used for preserving objects in cells. The different cells may be here mentioned; and it is recommended that these should always be kept some time before use in order that the cement may become perfectly dry; and care must be taken that no cement be used on which the preservative liquid employed has any action whatever.

Cement Cells.—Where the object is not very thick, this kind of cell is generally used. They are easily made with the turntable before described; but when the objects to be preserved are *very* minute, these cells need not be much deeper than the *ordinary* circle of cement on the slide. When, however, a comparatively great depth is required, it is sometimes necessary to make the wall of the cell as deep as possible, then allow it to dry and make another addition. Of these cements gold-size is one of the most trustworthy, and may be readily used for the shallow cells. The asphaltum and india-rubber, before noticed, I have found very durable when well baked, and exceedingly pleasant to work with. It may be used of such a thickness as to give space for tolerably large objects. Black japan also is much used. Many cements, however, which are recommended by some writers, are worse than useless, owing to the brittleness which renders their durability uncertain, as sealing-wax, varnish, liquid-glue, &c. Dr. Bastian says the best cement for liquid cells is one, much used in Germany, made by adding a considerable quantity of nitrate of bismuth to a solution of gum mastic in chloroform. It can be procured at almost any optician's.

The student may feel himself at a loss in choosing the cement which will give him the *safest* cells, many of them becoming partially or wholly dry in a year or two, as stated in another place. I can only give him a few general directions, and he must then use his own judgment. Of course it would be lost labour to employ any cement upon which the preservative liquid has *any action whatever*. It is also a good rule to avoid those in whose composition there are any particles which do not become a thorough and intimate portion, as these unreduced fragments will almost certainly, sooner or later, prepare a road by which the liquid will escape; and, lastly, whatever cement he uses, the cells are always better when they have been kept a short time before use.

GUTTA-PERCHA RINGS have been recommended by some, as affording every facility for the manufacture of cells for liquids; but they cannot be recommended, as, after a certain length of time, they become so brittle as to afford no safeguard against ordinary accidents. Some have also used india-rubber bands thickly coated with various varnishes; but these I consider less trust-worthy than gutta-percha, as they become thoroughly rotten in ordinary use after a short probation.

Often the cells must necessarily be of a large size, and for this reason are made by taking four strips of glass of the thickness and depth required, and grinding the places where these are to meet with emery, so as to form a slightly roughened but flat edge. The glass strip must also be ground on the side where it meets the plate, and each piece cemented with the marine-glue mentioned in Chapter II in the following manner:—On that part of the glass to which another piece is to be attached should be laid thin strips of the glue; both pieces must then be heated upon a small brass table, with the aid of the spirit-lamp, until the strips become melted; the small piece is then to be taken up and placed upon the spot to which it is to be attached, and so on until the cell is completed. It will be found necessary to spread the glue over the surface required with a needle or some other instrument, so that an unbroken line may be presented to the wall of the cell, and no bubbles formed. Too great a heat will burn the marine-glue, and render it brittle; care must be therefore taken to avoid this.

When shallow cells are required, those which are made by grinding a concavity in the middle of an ordinary slide will be found very convenient. The concavities are cut both circular and oblong; and the surface being flat, the cover is easily fastened upon it. These are now cheap, and are very safe as to leakage. It is a very great improvement, where it can be done, to turn a shallow ring outside the concavity of the slide, but close to it. This prevents the cement with which the cover is fastened from running in.

Circular cells with a flat bottom used to be made by drilling a hole through glass of the required thickness, and fixing this upon an ordinary slide with marine-glue; but the danger of breakage and the labour were so great that this method is seldom used now, and, indeed, the rings about to be mentioned do away with all necessity for it.

Glass Rings.—Where any depth is required, no method of making a cell for liquids is so convenient as the use of glass rings, which are now easily and cheaply procurable. They are made of almost every size and depth, and, except in very extraordinary cases, the necessity for building cells is completely obviated. These rings have both edges left rough, and consequently adhere very well to the slide, this adherence being generally accomplished by the aid of marine-glue, as before noticed with the glass cells. Gold-size has been occasionally used for this purpose; and the adherence, even with liquid in the cell, I have always found to be perfect. This method has the advantage of requiring no heat, but the gold-size must be perfectly dry, and the ring must have been fixed upon the slide some time before use. Canada balsam has also been used for the same purpose, but cannot be recommended, as, when it is perfectly dry, it becomes so brittle as to bear no shock to which the slide may be ordinarily exposed.

Iron Rings.—Many have worked with these, having taken care to varnish thoroughly before using with any preservative liquid; but they are always untrustworthy, as they can never be guaranteed against the action of some salt in the liquid used. They can be procured beautifully made, and for *dry* cells cannot be surpassed. Zinc and pure tin rings may also be procured, and are excellent, especially the latter.

Vulcanite.—This substance is a great favourite with some of our working microscopists, as it is very slightly influenced by change of temperature. But my own opinion is that a glass cell is the safest and most satisfactory receptacle for any object in liquid, and if carefully prepared will not deceive the operator.

These are the cells which are mostly used in this branch of microscopic mounting. The mode of using them, and the different treatment which certain objects require when intended to be preserved in the before-mentioned liquids, may now be inquired into.

I may mention, however, that this class of objects is looked upon by many with great mistrust, owing to the danger there is of bubbles arising in the cells after the mounting has been completed, even for years. I know some excellent microscopists who exclude all objects in cells and preservative liquids from their cabinets, because they say that eventually almost all become dry and worthless; and this is no matter of surprise, for many of them do really become so. Perhaps this is owing to the slides being sold before they could possibly be thoroughly dry. As to the air-bubbles, I shall have something to say presently.

We will now suppose the cell employed has been made by placing a glass ring upon the slide with marine-glue or gold-size, and is quite dry. Around the edge of the cleaned thin glass which is to cover it, I trace with a camel-hair pencil a ring of gold-size, and also around the edge of the cell to which it is to adhere. Dr. Carpenter objects to this, as rendering the later applications of the gold-size liable to "run in." All danger of this, however, is completely obviated by leaving the slide and cover for awhile until the cement becomes partially fixed, but still adhesive enough to perform its function (Chapter III.). With many slides this is not accomplished in less than twenty-four hours, even if left two or three days no injury whatever ensues; but with other kinds an hour is too long to leave the exposed cement, so that the operator must use his own discretion. It is not always necessary to size the edge of the *cover*, since perfect adhesion may in many cases be secured without it, and it is always best to use the least quantity of cement that will answer, as it will then be less likely to run in. The liquid required may be drawn up by the mouth into the pointed tube mentioned in Chapter II., and then transferred to the cell. In the various books of instruction, the object is now to be placed in the cell; this, however, I think a great mistake, as another process is absolutely necessary before we advance so far. The cell, full of liquid, must be placed under the receiver of an air-pump, and the air withdrawn. Almost immediately it will be perceived that the bottom and sides of the cell are covered with minute bubbles, which are formed by the air that is held in suspension by the liquid. The slide may now be removed, and the bubbles may require the aid of a needle or other point to displace them, so obstinately do they adhere to the surface of the glass. This process may then be repeated, and one cause, at least, of the appearance of bubbles in cells of liquid will be removed. The object to be mounted should also be soaked in one or two changes of the preservative liquid employed, and, during the soaking, be placed under the air-pump and exhausted. It may then be transferred to the cell, and will probably cause the liquid to overflow a little. The cover with the gold-size applied to the edge must then be carefully laid upon the cell, and slightly pressed down, so that all air-bubbles may be displaced. The two portions of gold-size will now be found to adhere wherever the liquid does not remain, although the whole ring may have been previously wet. The outer edge of the thin glass and cell must now be perfectly dried, and a coating of gold-size applied. When this is dry, the process must be repeated until the cement has body enough to protect the cell from all danger of leakage. When some preservative liquids are used, a scum is frequently found upon the surface when it is placed in the cell, and this must be removed immediately before the cover is laid upon it.

I believe this method to be perfectly secure against leakage when carefully performed; and some of my friends have told me that their experience (that of some years) has been equally satisfactory.

In using some of the particular kinds of preservative liquids, it will be found necessary to make a slight change in the manipulation. This will be best explained by mentioning a few objects, and the treatment they require.

For the preservation of the Mosses, Algæ, &c., Deane's compound is much used, and considered one of the best media. The specimen to be mounted should be immersed in the compound, which must be kept fluid by the vessel containing it being placed in hot water. In this state the whole should be submitted to the action of the air-pump, as it is not an easy matter to get rid of the bubbles which form in and around the objects. The cell and slide must be warmed; and heat will also be necessary to render the gelatine, &c., fluid enough to flow from the stock-bottle. The cell may then be filled with the compound, and the specimen immersed in it. A thin glass cover must then be warmed, or gently breathed upon, and gradually lowered upon the cell, taking care, as with all liquids, that no bubbles are formed by the operation. The cover may be fixed by the aid of gold-size, Japan, or any of the usual varnishes, care being taken, as before, that all the compound is removed from the parts to which the varnish is intended to adhere.

The glycerine jelly of Mr. Lawrance, before mentioned, requires almost a similar treatment. "The objects to be mounted in this medium should be immersed for some time in a mixture of equal parts of glycerine and dilute alcohol (six of water to one of alcohol). The bottle of glycerine jelly must be placed in a cup of hot water until liquefied, when it must be used like Canada balsam, except that it requires less heat. A ring of asphaltum varnish round the thin glass cover completes the mounting."

The Infusoria (see Chapter IV.) are sometimes preserved in liquid; but present many difficulties to the student. Different kinds require different treatment, and consequently it is well, when practicable, to mount similar objects in two or more liquids. Some are best preserved in a strong solution of chloride of calcium, others in Thwaites' liquid, whilst a few keep their colour most perfectly when in glycerine alone. There can be little doubt that light is the bleaching agent in most cases. Many of them, however, are so very transparent that they present but faint objects for ordinary observation. For this reason, however, they are sometimes dyed in solution of magenta or other colour, as elsewhere noticed. The Desmidiaceæ require somewhat similar treatment, and may be mentioned here. The solution of chloride of calcium has been strongly recommended; but no preservative liquid seems to be without some action upon them. Both of the above classes of objects should be mounted in shallow cells, so as to allow as high a microscopic power as possible to be used with them.

Entomostraca.—In every ditch or place where vegetable matter exists, these little active, jerking pieces of life are certain to be found. They are covered with a horny transparent shell, and are various in form. Mr. Tatem gives the following, as the best way of preserving them:—When caught transfer to filtered water in watch-glasses for twenty-four hours, in order that the contents of the laden intestine may be discharged. Draw off the water and add a little spirit of wine, which quickly destroys life. Remove all dirt by aid of a camel-hair pencil, and place in a few drops of the medium used and water (half of each) until saturation is complete, and then put up in the medium in shallow cells. The medium advised is Mr. Farrant's, which will be found amongst those recommended.

Many of the ZOOPHYTES which are obtained on our sea-coasts are well preserved by mounting in cells, in the manner before mentioned, with Goadby's fluid, or distilled water with one of the additions noticed amongst the preservative liquids. For examination by polarized light, however, they are usually mounted in balsam (see Chapter IV.), whilst those in cells present a more natural appearance as to position, &c., for common study. The POLYZOA, also, are exquisitely beautiful objects for the microscope, but require some little care. They should be kept in sea-water until their tentacula are expanded, and may then be readily killed by plunging in cold fresh water. Thus all their beauty will be preserved, and they may be then mounted in one of the preservative liquids. Many operators speak well of distilled water well shaken with a few drops of creosote, as before mentioned.

As to the use of preservative liquids with the Diatomaceæ there are various opinions. Some experienced microscopists say that there is little or no satisfaction in mounting them in this way. Dr. Carpenter, however, explains this difference by his instructions as to what method should be used when certain ends are desired. He says: "If they can be obtained quite fresh, and it be desired that they should exhibit as closely as possible the appearance presented by the living plants, they should be put up in distilled water within cement cells; but if they are not thus mounted within a short time after they have been gathered, about a sixth part of alcohol should be added to the water. If it be desired to exhibit the stipitate forms in their natural parasitism upon other aquatic plants, the entire mass may be mounted in Deane's gelatine in a deeper cell; and such a preparation is a very beautiful object for black-ground illumination. If, on the other hand, the minute structure of the silicious envelopes is the feature to be brought into view, the fresh diatoms must be boiled in nitric or hydrochloric acid" (which process is fully described in Chapter III.). It is very convenient to have many of these objects mounted by two or more of the above methods; and if they are to be studied, this is indispensable. Mr. Hepworth once showed me about one hundred slides which he had mounted in various ways, for no other purpose than the study of the fly's foot.

My friend, Mr. Rylands, successfully mounts the diatoms in the state in which he finds them, and gave me the following method as that which he always employs. He says that he has had no failures, and hitherto has found his specimens unchanged. Take a shallow ring cell of asphalt or black varnish (which must be at least three weeks old), and on the cell, whilst revolving, add a ring of benzole and gold-size mixed in equal proportions. In a minute or two pure distilled water is put in the cell until the surface is slightly convex. The object having been already floated on to the cover (the vessel used for this purpose being an ordinary indian-ink pallet), is now inverted and laid carefully upon the water in the cell. By these means the object may be laid down without being removed. The superfluous moisture must not be ejected by pressure, but a wetted camel-hair pencil, the size made in an ordinary quill, being partially dried by drawing through the lips, must be used repeatedly to absorb it, which the pencil will draw by capillary attraction as it is very slowly turned round. When the cover comes in contact with the benzole and gold-size ring, there is no longer any fear of the object being removed, and a slight pressure with the end of the cedar stick of the pencil will render the adhesion complete, and cement the cover closely and firmly to the cell. When dry, an outer ring of asphalt makes the mounting neat and complete.

The *Fungi* have been before mentioned; but it may be here stated that some few of the minute forms are best preserved in a very shallow cell of liquid. For this purpose creosote-water may be advantageously used.

The *antennæ* of insects have been before noticed as being very beautiful when mounted in balsam. This is readily accomplished when they are large; but those of the most minute insects are much more difficult to deal with, and are less liable to injury when put up in fluid. *Goadby's Fluid* serves this purpose very well; but, of course, the object must be thoroughly steeped in the liquid before it is mounted, for a longer or shorter time according to the thickness.

The *eggs of insects* afford some worthy objects for the microscope, amongst which may be mentioned those of the common cabbage butterflies (small and great), the meadow-brown, the puss-moth, the tortoiseshell butterfly, the bug, the cow-dung fly, &c. These, however, shrivel up on becoming dry, and must, therefore, be preserved in some of the fluids before mentioned. To accomplish this no particular directions are required; but the soaking in the liquid about to be employed, &c., must be attended to as with other objects.

Glycerine may be advantageously used for the preservation of various insects. These should first be cleaned with alcohol to get rid of all extraneous matter, and then, after soaking in glycerine, be mounted with it like other objects. There is, however, a difficulty in clearing glycerine from the edge of the thin glass cover; but Mr. Whalley told me he met with no annoyance. After laying the cover upon the object with the glycerine, he took away all the superfluous liquid with a small piece of linen, cleaning it at last with a *damped* piece of the same. The small quantity of water which gets mingled with the glycerine does no injury, and the edges can be thus cleaned perfectly enough for any cement to adhere. Mr. Suffolk, at the Quekett Club, said:—When the cell was closed he varnished it with a coating of common liquid-glue, and when this was dry he put it under the tap and thoroughly washed it, in order to remove any glycerine which might remain outside. After carefully drying the slide with blotting-paper, he gave it another coating of the liquid-glue, and when dry repeated the washing process, and after having given it a third coating of liquid-glue in the same manner, he gave it a final coat of gold-size, and he never had any trouble with cells closed in this manner. Mr. Hislop, at the same place, said:—His plan was, to make a good seat for the cover first by a thick ring of gum dammar—allow this to become sticky; next put in the glycerine, lay on the cover, and then carefully wash off all superfluous glycerine. When perfectly well-washed and dried lay on two or three coats of gum dammar to finish it.

Some insects, such as May-flies, &c., are, however, often preserved by immersion in a solution of one part of chloride of calcium in three or four parts of water; but this has not been recommended amongst the preservative liquids, as the colour, which is often an attractive quality of this class of objects, is thereby destroyed.

We have now noticed the treatment which must be applied to those objects which are to be preserved in liquids and cells. We may here state that all slides of this kind should be examined at short intervals, as they will be found now and then to require another coating of varnish round the edge of the thin glass cover to prevent all danger of leakage. The use of the air-pump, in the first instance (as before recommended), and this precaution as to the varnish, will render the slides less liable to leakage and air-bubbles, which so very frequently render them almost worthless.

CHAPTER VI.

SECTIONS AND HOW TO CUT THEM, WITH SOME REMARKS ON DISSECTION.*

MANY objects are almost worthless to the microscopist until extraneous matter is removed from them; and this is frequently difficult in the extreme to perform satisfactorily. As an instance, certain Foraminifera may be mentioned in which the cells are placed one upon another, consequently the object must be reduced to a certain degree of *thinness* before a single uniform layer of these cells can be obtained to show something of the internal arrangements.

Most animal and vegetable forms require an examination of the separate parts before much can be known about them. The mass must be divided into separate portions, each part intended to be preserved being cleaned from the useless matter with which it is surrounded. It will frequently be found necessary to make thin sections, which from a very tender substance is no easy matter; and much patience will be necessary to attain anything like proficiency.

This making of sections was not until very recently undertaken by many except those belonging to the medical profession, but I do not see why this should be so, as much may be accomplished by a persevering and interested mind where there is time for entering into the subject. I will therefore make an attempt to give some instructions on this subject also. We will first consider the cutting of sections from hard substances, in which the ordinary knife, chisel,

* As some knowledge of dissection is necessary to success in injection, additional matter on this subject will be found in Chapter VII.

&c., are of no avail. Most of these require no particular care in mounting, but are placed in balsam like the other objects noticed in Chapter IV.: where, however, any special treatment is necessary it will be commented upon as we proceed.

SHELLS, &c.—It is seldom, if ever, necessary to possess apparatus for this process except a small thin saw made with a steel blade, for which a piece of watch-spring serves very well; a fine stone such as is used for sharpening penknives; and two smooth leather strops, one of which is to be used with putty-powder to polish the section after grinding, and the other dry, to give the final surface. It is, however, very convenient to have three or four files of different degrees of fineness. A very useful implement in this process is the Corundum file or rubber, sold by most dealers in watchmakers' tools. It may be procured of almost any size or grain, either circular or flat, and will cut almost anything. They possess the very great advantage of not carrying much, if any, impurity into the texture of the object upon which they are used. The shell, if very thick, may be divided by using the watch-spring saw; and this section may then with ordinary care be rubbed down with water on the stone until one side of it is perfectly flat. When this is accomplished it must be rubbed with putty-powder upon the strop, and finally upon the other strop without the powder. This surface will then be finished, and must be firmly united to the slide in the position it is intended to occupy. To do this a small quantity of Canada balsam may be dropped upon the middle of the slide and heated over the lamp until on cooling it becomes hard; but this must be stopped before it is rendered brittle. Upon this the polished surface must be laid, and sufficient heat applied to allow the object to fall closely upon the slide, when slight pressure may be used to force aside all bubbles, &c. On cooling, the adherence will be complete enough to allow the same grinding and polishing upon the upper surface which the lower received. Whilst undergoing this, the section must be examined from time to time to ascertain whether the necessary degree of thinness has been reached. When this is the case the section should be washed thoroughly and dried. It must then be covered, which is best done by using ordinary Canada balsam, as recommended in Chapter IV.; or, if the section is to be mounted dry, it must be freed from balsam by washing, or soaking if necessary, in turpentine or other solvents.

Sections of some exquisitely beautiful objects are cut with much less trouble than the above. The Orbitolite, for instance, may be prepared in this manner. Take the object and by pressure with the finger rub the side upon a flat and smooth sharpening stone with water until the portion is reached which it is wished to show. The strength of the object will easily allow this to be accomplished with ordinary care. This side may then be attached to the glass slide with heated balsam, as above described, and the object may then be gently rubbed down to the degree of thinness required to show it to the best advantage. After removing all disengaged matter from the object by washing and thoroughly drying, it may be mounted in balsam in the usual manner, when it is equally beautiful as a transparent or opaque object. From this it will be seen that in many instances where a smooth stone is found sufficient for the work (which is often the case when the section is mounted in balsam) the final process of polishing advised above may be dispensed with, as in the Orbitolite, Nummulite, &c., &c. It is quite necessary that the stones on which the objects are rubbed be *perfectly flat*, otherwise one side must be acted upon before the other, and it will be found impossible to attain anything like uniformity. Where it is not practicable to cut a section, and the object is very thick, a coarse stone may be first used to reduce it and the smoother afterwards.

The consideration of the cutting of sections from shells would scarcely be deemed complete without some mention of what Dr. Carpenter terms the decalcifying process. Muriatic acid is diluted with twenty times its volume of water, and in this the shell is immersed. After a period, differing according to the thickness of the shell, the carbonate of lime will be dissolved away, and a peculiar membrane left, showing the structure of the shell very perfectly. This may be mounted dry, in balsam, or sometimes in liquid, according to the appearance of the object; but no rule can be given. The discretion of the student, however, will enable him to choose the most suitable method.

From some shells it is easy to divide thin plates, or laminæ, which require nothing but mounting in Canada balsam to show the texture very well. In working, however, with those which are pearly, it will be found that experience and patience are needed, as they are very brittle and peculiarly hard; but a little practice will overcome these difficulties.

Amongst the Echinodermata, which include the star-fishes, sea-hedgehogs, &c., there are many whose outer surface is covered with spines, or thin projections. Some of these are sharp and thorn-like, others blunt, longer or shorter, and, indeed, of endless variety. In many of these, when a section is made, rings are seen which have a common centre, with radiating supports, resembling sections of some of the woods. These are very beautiful objects, and methods of procuring them may now be considered. It is the best to cut as thin a section as can safely be got with the watch-spring saw first, when the smooth sharpening stone may be used to polish one side, which is easily accomplished with water only. When this is effected, it must be washed clean, and *thoroughly* dried, and then may be united to the slide in the same manner as before recommended for the Orbitolite, &c. If it is ever necessary to displace it on account of inequalities, bubbles, or other remediable fault, this may be done by warming the slide; though too much heat must be avoided, otherwise fresh bubbles will certainly be produced. The covering with thin glass, balsam, &c., will present no difficulty to the student; but he must remember that the transparency is somewhat increased by this last operation.

Corals are often treated in this way, in order to reveal their structure. Except, however, the student has had much practice, he will often find this a most difficult task, as many of them are exceedingly brittle and hard. He will find the method before described equally applicable here, and should take both horizontal and vertical sections.

COAL.—This substance is one of the most interesting objects to the microscopist. It is, of course, of vegetable origin; and though it is in many cases in such minute separate portions as to have lost all appearance of vegetation, yet it is very frequently met with in masses, bearing the form, even to the minute markings, of wood, in various directions. To see this and prepare it for microscopic research, a suitable piece of coal must be obtained; but in every case the cutting and preparation of these sections require great care and skill. Sometimes the coal is first made smooth on one side, fastened to the glass, reduced to the requisite degree of thinness, and finished in the method before described. This mode of treating it is sometimes, however, very tantalizing, as, at the last moment, when the section is about thin enough, it often breaks up, and so renders the trouble bestowed upon it fruitless. The dark colour and opacity of coal render an extraordinary thinness necessary, and so increase the liability to this accident.

Mr. Slade recommends that the piece of coal, having been smoothed on one side, be cemented on that side to a glass slip by marine-glue of the best quality, quite free from undissolved or foreign matter. Great care must be taken to press out all air-bubbles, the coal breaking up at such places as it gets thin, a hole resulting. It may then be reduced in the usual way, and when thin enough mounted in Canada balsam and covered by thin glass.

Perhaps the best method which can be pursued is that recommended in the Micrographic Dictionary, which is as follows:—"The coal is macerated for about a week in a solution of carbonate of potash; at the end of that time it is possible to cut tolerably thin slices with a razor. These slices are then placed in a watch-glass with strong nitric acid, covered, and gently heated; they soon turn brownish, then yellow, when the process must be arrested by dropping the whole into a saucer of cold water, else the coal would be dissolved. The slices thus treated appear of a darkish amber colour, very transparent, and exhibit the structure, when existing, most clearly. We have obtained longitudinal and transverse sections of coniferous wood from various coals in this way. The specimens are best preserved in *glycerine* in cells; we find that spirit renders them opaque, and even Canada balsam has the same defect. Schultze states that he has brought out the cellulose reaction with iodine in coal treated with nitric acid and chlorate of potash." Now and then in coal we meet with a half-formed carbon-looking substance which is no more difficult to work with than ordinary charcoal. From this it is an easy thing to procure interesting slides.

Cannel-coal is so close and firm in its structure as to be much used instead of jet in the manufacture of ornaments: it takes a beautiful polish, and consequently presents the student with none but ordinary difficulties in getting sections of it. Its formation is somewhat different from that of coal, sometimes showing the transition very clearly.

Fossil Wood.—This is very often brittle and requires great care in cutting. There are, however, different kinds of fossil wood, but to obtain anything like certainty and perform much work a lathe is necessary. I know of no method better than that given by Mr. Butterworth, and shall therefore make use of his words. First, I will begin with the cutting. To the framework of an ordinary foot-lathe I attach an upright spindle (see engraving). On this upright spindle I drive by a band passing over "carry-pulleys" from the wheel below. On the top of this spindle I fix my cutting-disc, which is made from a very thin piece of sheet iron, and is about six inches in diameter.

The edge of this saw I charge with diamond-powder. To the edge of the saw I hold my specimen, and as it cuts I lubricate the edge with a small brush dipped in turpentine.

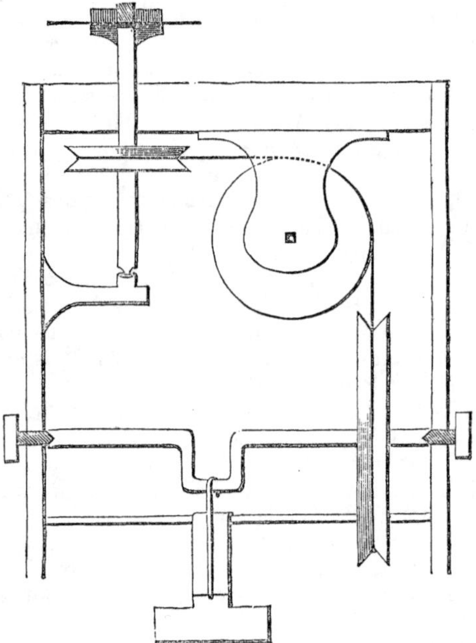

With this method I have cut sections of fossil wood so thin that all its structure has been well defined and required nothing but mounting in balsam; this has been *silicated* fossil wood. In cutting *calcareous* fossil wood, I have to cut the sections thicker and grind them down. My grinding apparatus is composed of leaden laps, which I make to revolve in a horizontal position on the same upright spindle on which I fix my cutting saw. I use two laps, one for rough grinding, the other for smoothing. I use No. 1. emery and a little water with the first, and flour of emery with plenty of water on the second. In preparing a specimen, I first grind a smooth surface on one side, and then fix it to a plate of glass, of such a size as will suit my specimen, with Canada balsam. I then reduce it in thickness on the rough lap till I begin to see the light through it. Then I begin with the smoothing lap, and reduce it with flour of emery until every part of its structure is distinct. If I choose to polish the specimen I do so on a lap made of plush cloth or cotton vevet and putty-powder. I then float them off the slide on which they have been ground, and fix them on another with Canada balsam. I prefer, where it is practicable, to mount them in balsam under a thin cover in the usual way, as I am satisfied that the structure is better brought out.

In *flint* there are often found remains of sponges, shells, Diatomaceæ, &c.; but to show these well, sections must be cut and polished by the lathe and wheel of the lapidary, which the microscopic student seldom possesses. Thin chippings may, however, be made, which when steeped in turpentine and mounted in balsam, will frequently show these remains very well.

Teeth are very interesting objects to all microscopists, more especially to those who give much study to them; as the class of animal may very frequently be known from one solitary remaining tooth. To examine them thoroughly, it is necessary to cut sections of them; but this is rather difficult to perform well, and needs some experience. Some instructions, however, will at least lessen these difficulties, and we will now endeavour to give them.

Sections of teeth and bone may be successfully made by rubbing slices cut with a saw between two plates of ground glass, with water and a little powdered pumice-stone, the old and partially worn glass being kept for the final polishing of the sections.

It is generally thought that Canada balsam injures the finer markings of these sections, consequently, they are almost invariably mounted *dry*. A thin piece is first cut from the tooth with the saw of watch-spring before mentioned, if possible; but should the substance be too hard for this, the wheel and lathe must be used with diamond dust. If this cannot be procured, there is no alternative but to rub down the whole substance as thin as practicable on some coarse stone or file, or best of all the corundum rubber. The surface will then be rough; but this may be much reduced by rubbing upon a flat sharpening stone with the finger, or a small piece of gutta-percha upon the object to keep it in contact. The scratches may be much lessened by this, but not so thoroughly removed as microscopic examination requires in dry sections. It must, therefore, be polished with the putty-powder and dry strop, as recommended in the working of shell-sections. The other side of the section of the tooth may then be rubbed down to the requisite thinness, and polished in the same manner, when the dust and other impurities must be removed by washing, after which the section must be carefully dried and mounted. Sometimes it may be deemed desirable to make a preparation of the teeth *in situ*; for this purpose take the *lower* jaw of some animal like the rat, weasel, or guinea-pig, and soak it in absolute alcohol first, let that evaporate out, then soak in the solution of balsam and benzole; when that has evaporated to hardness, grind down the jaw as a section, the teeth are fixed in by the balsam. Some of these sections are equally interesting as opaque or transparent objects.

The dentine of the teeth may be decalcified by immersion of the section in dilute muriatic acid; after drying and mounting in Canada balsam it presents a new and interesting appearance, showing the enamel fibres very beautifully when magnified about three hundred diameters. A friend tells me that after submersion of the *whole tooth* in the acid he has been able to cut sections with a razor.

Sections of Bone.—With the aid of the microscope few fragmentary remains have proved so useful to the geologist and students of the fossil kingdom as these. From a single specimen many of our naturalists can tell with certainty to what *class* of animal it has once belonged. To arrive at this point of knowledge much study is necessary, and sections of various kinds should be cut in such a manner as will best exhibit the peculiarities of formation. The methods of accomplishing this will now be considered. It may, however, be first mentioned that the chippings of some bones will be found useful now and then, as before stated with flint, though this is by no means a satisfactory way of proceeding. Sometimes the bones may be procured naturally so thin that they may be examined without any cutting; and only require mounting *dry*, or in *fluid*, as may be found the best.

When commencing operations we must provide the same apparatus as is needed in cutting sections of teeth, before described. A fine saw, like those used for cutting brass, &c., two or three flat files of different degrees of coarseness; two flat "sharpening" stones; and a leather strop with putty-powder for polishing. As thin a section as possible should first be cut from the part required by the aid of the fine saw; and it is better when in this state to soak it for some short time in camphine, ether, or some other spirit to free it from all grease. With the aid of a file we may now reduce it almost to the necessary degree of thinness, and proceed as before recommended with teeth. The "sharpening" stone will remove all scratches and marks sufficiently to allow it to be examined with the microscope to see if it is ground thin enough; and if it is to be mounted *dry* we must polish it with putty-powder and water upon the strop to as high a degree as possible, and having washed all remains of polishing powder, &c., from the section we must place it upon the slide and finish it as described in Chapter III. But where these sections are required for mounting in balsam a less amount of *polish* is necessary; thus rendering the whole process much more readily completed.

If the bone is not sufficiently hard in its nature to bear the above method of handling whilst grinding and polishing—as some are far more brittle than others—as thin a section as possible must first be cut with the saw, and one surface ground and polished. The piece must then be dried and united to the glass by heated balsam in the same manner as shells, &c. After which the superabundance of balsam must be removed from the glass; then rub down upon the stone and strop as before. Great care must be taken that the canals be not filled during the process with the dust of the bone, or of the polishing material. Dr. Beale, in the journal of the Q. M. C. takes occasion to say "that he cannot admit that the best way of preparing such sections is by grinding down, since it is too liable to fill the canals with *débris*." He recommends that a fresh bone be taken—and a small slice cut off by a strong sharp knife. This is then to be immersed in carmine dissolved in ammonia—the ammonia being first neutralized by acetic acid. The walls of the vessels which penetrate the *lacunæ* and *canaliculi* are by this means stained crimson, and thus the true structure of bone is rendered visible. When the polishing is completed the whole slide must be immersed in chloroform, ether, or some other spirit, to release and cleanse the section, when it may be mounted as the one above mentioned.

Some have recommended a strong solution of isinglass to affix the half-ground teeth or bones to the glass as causing them to adhere very firmly and requiring no heat, and also being readily detached when finished.

The reason why the sections of bone are usually mounted *dry* is that the *lacunæ*, bone cells, and *canaliculi* (resembling minute canals) show their forms, &c., very perfectly in this state, as they are hollow and contain air, whereas if they become filled with liquid or balsam—which does sometimes occur—they become almost indistinguishable. There are some dark specimens, however, where the cells are already filled with other matter, and it is well to mount these with balsam and so gain a greater degree of transparency.

To obtain a true knowledge of the structure of bone, sections must be cut as in wood, both transversely and longitudinally; but with *fossil* bones, without the lapidary's wheel, it is a laborious task, and indeed can seldom be properly accomplished. In this place, also, it may be mentioned that by submitting bone to the action of muriatic acid diluted ten or fifteen times with water, the lime, &c., is dissolved away and the cartilage is left, which may be cut into sections: in *caustic potash* the animal matter is got rid of. Both of these preparations may be mounted in fluid.

The method of cutting thin sections of bone may be also employed with the stones of fruit, vegetable ivory, and such like substances; many of which show a most interesting arrangement of cells, especially when the sections are transverse. Most of these objects present a different appearance when mounted *dry* to that which they bear when *in balsam*, owing to the cells becoming filled; and to arrive at a true knowledge of them we must have a specimen mounted in both ways.

Some will perhaps remark that most of the directions for section cutting are given to those who are totally without artificial power, and must rely upon their own manual exertions. I reply that these hints are mostly given to such; but Mr. Butterworth's directions to use the *lathe* are so ample, that a repetition of them at the mention of each class of "sectional" substance would be mere tautology.

To those who study polarized light, few objects are more beautiful than sections of the different kinds of horn. We will briefly inquire into the best method of cutting these. There are three kinds of horn, the first of which is hard, as the stag's, and must be cut in the same manner as bone. The second is somewhat softer, as the cow's. The third is another and still softer formation, as the "horn" (as it is termed) of the rhinoceros. In cutting sections of the two last we should succeed best by using the machine invented for these purposes, which I shall shortly describe when the method of cutting wood is considered. To aid us in this when the horn is hard it must be boiled for a short time in water, after which the cutting will be more easily affected. The sections should be both transverse and longitudinal, those of the former often showing cells with beautiful crosses, the colours with the selenite plate being truly splendid. Of this class the rhinoceros horn is one of the best; but the buffalo also affords a very handsome object. The cow's, and indeed almost every different kind of horn, well deserves the trouble of mounting. Whalebone, when cut transversely, strongly resembles those of the third and softer formation. All these are best seen when mounted in Canada balsam, but care must be taken that they have been thoroughly dried after cutting, and then steeped in turpentine.

An interesting object may also be procured from whalebone by cutting long sections of the hairs of which it is composed. Down the centre of each hair we shall find a line of cells divided from one another very distinctly. And (as recommended in the Micrographic Dictionary) if whalebone be macerated twenty-four hours in a solution of caustic potash it will be softened, and by afterwards digesting in water, the outer part will be resolved into numerous transparent cells, which will show more plainly the structure of this curious substance.

An object which frequently comes to the hand of any man who moves about in the world is a porcupine quill. This is a really valuable object for the microscopist. Transverse and longitudinal sections possess their respective beauty; and their appearance varies somewhat as to the distance from the point at which the section is made. Soaking in hot water for a short time renders it easy enough to cut, and when dry and mounted in balsam the student is well repaid.

In a former chapter, hairs were mentioned, their many and interesting forms, and their beauty when used with polarized light. The sections of them, however, are no less a matter of study, as this mode of treatment opens to sight the outer "casing," and the inner substance somewhat resembling the pith of plants.

It would be out of place to enter into the description of the different forms met with; but the ways in which sections are to be procured may be noticed. If transverse sections are required, some place a quantity of hairs betwixt two flat pieces of cork, which by pressure hold them firmly enough together to allow the required portions to be cut with a razor. Others take a bundle of the hairs and dip it into gum or glue, which gives it when dry a solidity equal to wood. Sections of this are then cut with the machine mentioned a little further on, and these may be mounted in balsam. The human hair is easily procured in the desired sections by shaving as closely as possible a second time and cleansing from the lather, &c., by carefully washing. Most hairs, however, should be examined both transversely and longitudinally. It is not difficult to procure the latter, as we may generally split them with the aid of a sharp razor. In a great number of hairs there is a quantity of greasy matter which must be removed by soaking in ether or some other solvent before mounting.

We may next consider the best method of procuring *sections of wood*, which must be cut of such a degree of thinness as to form transparent objects, and so display all the secrets of their structure. There is no monotony in this study, as the forms are so various, and the arrangement of the cells and woody fibre so different, that the microscopist may find endless amusement or study in it. From a single section the *class* of trees to which it has belonged may be known, often even when the wood is *fossil*. The apparatus best adapted for cutting these sections is made as follows:—A flat piece of hard wood, about six inches long, four wide, and one thick, is chosen, to which another of the same size is firmly fixed, so as to form, in a side view, the letter T. On one end of the upper surface is fastened a brass plate, perfectly flat, in the centre of which a circular opening is cut about half an inch in diameter.

Coinciding with this opening is a brass tube, fixed in the under side of the table (if it may be termed so). This tube is so cut at the bottom as to take a fine screw. Another screw is also placed at the same end of the "table," which works at right angles to this, so that any substance in the tube may be wedged firmly by working this last screw. To use this instrument, the piece of wood or other object of which a section is required must be placed in the tube, when, by turning the screw underneath, the wood is raised above the brass plate more or less as wished, and by using the screw at the end, it is held firmly in the same position. With a flat chisel the portion of the object which projects above the surface of the brass plate may now be cut off, and by means of the bottom screw another portion may be raised and treated in the same manner. As to the thickness of which objects should be cut, no proper directions can be given, as this differs so greatly that nothing but experience can be any guide. The same thickness can be obtained by working the screw underneath in uniform degrees, the head being marked for this purpose; and where the substance to be cut is *very* much smaller than the hole in the brass plate, it may be wedged with cork.

*As this instrument is peculiarly adapted for cutting wood (though used for other substances, as before mentioned), I shall notice a few particulars concerning this branch of sections. It may here be remarked, that to obtain anything like a true knowledge of the nature of wood, it should be cut and examined in at least two directions, *across* and *along*. The piece of wood is often placed in spirits for a day or two, so that all resinous matter may be dissolved out of it; it must then be soaked in water for

* M. Mouchet, in order to avoid all danger of "beards" in cutting wood sections, procured a knife with a semicircular blade. This was fastened at the end upon a flat plate, in order to revolve, as we may call it, the handle being long enough to give leverage for any required power. The wood supporter being placed in a favourable position, the knife is easily brought round, and the section cut by a circular action.

the same length of time, so as to soften and render it easy to cut. Sections may then be obtained in the manner just described, but they often curl to such a degree from their previous immersion in water as to render pressure necessary to flatten them until dry. They are often mounted *dry*, and require no care beyond other objects, as in Chapter III. Some, however, are best mounted in balsam, particularly the long sections when used for the polariscope; these must be soaked in turpentine, and the greatest care taken that all air-bubbles are removed. Others are thought to be most useful when mounted in shallow cells with some of the *preservative liquids* mentioned in Chapter V.—weak spirit and water, chloride of calcium solution of the strength of one part of the salt to three parts of distilled water, &c.

The above "*section-cutter*" may not be within the reach of every student, nor is it absolutely necessary; though where any great *number* of specimens is required it is very useful, and insures greater uniformity in the thickness. Many employ a razor for the purpose, which must always be kept sharp by frequent stropping. Sections of leaves also may be procured by the same means, though, as before mentioned, they are sometimes divided by stripping the coatings off with the fingers. The cells which come to sight by cutting some of the orchideous plants are most interesting. To cut these leaves they may be laid upon a flat piece of cork, thus exposing the razor to no danger of injury by coming in contact with the support. It may be mentioned here that the *razor* may also be used in cutting sections of the rush, than which a more beautiful object can scarcely be found when viewed transversely, as it shows the stellate arrangements of the parenchyma. This should be mounted *dry*. In the same way sections of the leaf-stalks of ferns may also be cut, some of which, as Dr. Carpenter states, show the curious ducts very beautifully, especially when cut rather obliquely.

It has been found a ready method of cutting sections of the rush and such like plants, to suck a solution of gum up into the pith, and when this is dry thin sections could be cut and the gum washed out again, and these could be mounted in balsam.

The plan adopted by most practical histologists for cutting sections of soft tissues is as follows :—The tissue to be cut is first hardened by immersion in a chromic acid solution varying in strength from 0·25 to 2 per cent., or by immersion in alcohol. The substance to be cut may then be embedded in melted wax and spermaceti, in proportions suitable to the nature of the substance to be cut; when this is cold the section may be cut with a razor ground flat on one side, and may then be floated off in spirits of wine.

These sections mount very well in Canada balsam, if after being removed from the spirit they are immersed in oil of cloves till they become clear, then put into turpentine before the balsam. The thinness of the section will depend very much on the dexterity of the operator, but section-cutting instruments for soft tissues can now be obtained at most scientific instrument shops.

When sections of softer substances are required, no instrument can be compared with "Valentin's knife," which consists of two steel blades lying parallel with each other and attached at the lower end. The distance of separation may be regulated at will by a small screw near the handle. When, therefore, a section is wanted, the substance must be cut through, and betwixt the blades a thin strip will be found, which may be made of any thickness, according to the distance of their separation. By loosening the screw the blades may be extended, and the section may be floated out in water if the damp will not injure it. The knife cuts much better if dipped in water or glycerine immediately before use, and also when the substance to be operated upon is wet, or even under water altogether; but care must be taken, after use, to clean the blades thoroughly and oil them before laying by, if the place is at all damp. This instrument is most useful in such subjects as anatomical preparations where the sections are required to show the position of the different vessels, &c.; but, as before stated, is very valuable for all soft substances. As an instance of this, it may be mentioned, that it is frequently used in cutting sections of sponges; but as these are often very full of spicula, it is much better to press the sponge flat until dry, and then cut off thin shavings with a very sharp knife; these shavings will expand when placed in water. After this they may be laid betwixt two flat surfaces and dried, when they may be mounted as other dry objects, or, when desirable, in balsam.

Valentin's knife is very much used in taking sections of skin, which are afterwards treated with potash solution, acids, &c., to bring out in the best way the different portions. Dr. Lister's mode, however, of getting these is thus given in the *Microscopic Journal*:—"But I afterwards found that much better sections could be obtained from dried specimens. A portion of shaved scalp being placed between two thin slips of deal, a piece of string is tied round them so as to exercise a slight degree of compression; the preparation is now laid aside for twenty-four hours, when it is found to be dried to an almost horny condition. It then adheres firmly by its lower surface to one of the slips, and thus it can be held securely, while extremely thin and equable sections are cut with great facility in any plane that may be desired. These sections, when moistened with a drop of water and treated with acetic acid, are as well suited for the investigation of the muscular tissue as if they had not been dried."

There are many who almost confine their attention to polarized light and its beautiful effects. Such would not deem these efforts to aid the student in cutting sections complete, without some notice of those which are taken from various crystals, in order to display that curious and beautiful phenomenon, *the rings with a cross*. The arrangement of these is somewhat changed by the crystal which affords the section; but nitrate of potash gives two sets of rings with a cross, the long line of which passes through both, the short line dividing it in the middle.

The process of cutting these sections is rather difficult, but a little care and perseverance will conquer all this. The following is extracted from the *Encyclopædia Metropolitana:* "Nitre crystallizes in long six-sided prisms whose section, perpendicular to their sides, is the regular hexagon. They are generally very much interrupted in their structure; but by turning over a considerable quantity of the ordinary saltpetre* of the shops specimens are readily found which have perfectly transparent portions of some extent. Selecting one of these, cut it with a knife into a plate *above* a quarter of an inch thick, directly across the axis of the prism, and then grind it down on a broad wet file till it is reduced to about one quarter or a sixth of an inch thick, smooth the surface on a wet piece of emeried glass, and polish on a piece of silk strained very tight over a strip of plate-glass, and rubbed with a mixture of tallow and colcothar of vitriol. This operation requires practice. It cannot be effected unless the nitre be applied wet and rubbed till quite dry, increasing the rapidity of the friction as the moisture evaporates. It must be performed in gloves, as the vapour from the fingers, as well as the slightest breath, dims the polished surface effectually. With these precautions a perfect vitreous polish is easily obtained. We may here remark, that hardly any two salts can be polished by the same process. Thus, Rochelle salt must be finished wet on the silk, and instantly transferred to soft bibulous linen and rapidly rubbed dry. Experience alone can teach these peculiarities, and it is necessary to resort to contrivances (sometimes very strange ones) for the purpose of obtaining good polished sections of soft crystals, especially of those easily soluble in water.

* Sometimes the saltpetre of the shops is nitrate of *soda*, and as this is slightly deliquescent, it is well to be certain that we have the nitrate of *potash*, which is free from this defect.

"The nitre is thus polished on both its surfaces, which should be brought as near as possible to parallelism."

Some sections of the naturally formed crystals also show the "rings" very well,—as Iceland spar, which gives a single ring and cross; but the difficulty of cutting and polishing them is almost too great for the amateur, and must be left to the lapidary. This curious phenomenon, however, may be seen by using a plate of ice uninterruptedly formed of about one inch in thickness.

Before concluding these remarks on sections, I must mention a few difficulties which may be met with, and their remedies. The foremost of these is the softness of some objects, which have not resistance enough in themselves to bear cutting even with the sharpest instruments. This may often be removed by soaking in a solution of gum, and then drying, which will render the substance firm enough to be cut, when the sections must be steeped in water, and the gum thus removed. Small seeds, &c., may be placed in wax when warmed, and will be held firmly enough when it is again cold to allow of them being cut into sections.* And, lastly, where a substitute for a microscopist's hand-vice is required, a cork which fits any tube large enough may be taken and split, the object being then placed between the two parts, and the cork thrust into the tube, a sufficient degree of firmness will be obtained to resist any necessary cutting.

* Mr. T. K. Parker informs me that he uses paraffine as an "object-support" when sections are required, as follows:—"The mixture I use for embedding objects consists of solid paraffine (ordinary paraffine candles will do very well) melted down and mixed with a little paraffine oil, without which the paraffine is too hard to be easily cut. The mixture when cold is cut into suitable pieces, a hole is scooped out in the centre, the object to be cut placed in it, and a little of the melted mixture poured round it. The sections are cut with an ordinary razor, which, as well as the object, must be continually wetted with spirit. This method is useful for all objects which are either too small for the hand or too soft or brittle to be cut in the ordinary way. It is especially useful for histological specimens, leaves, embryos, &c."

The ether process of drying tissues has been described by Mr. Suffolk, at a meeting of the Quekett Microscopical Club, and was communicated by Mr. Crooker to him—it is as follows :—A wide-mouthed well-stoppered bottle must be selected. At the bottom is placed a slice from the bowl of a tobacco-pipe, forming a support for a Berlin crucible with its cover. A quantity of fused chloride of calcium in fragments is placed at the bottom of the bottle, which is nearly filled with pure ether, so that the crucible may be covered. The tissue to be dried is placed in the crucible, and is covered, if necessary to keep it from floating, by a piece of glass. The ether takes water from the tissue, and the chloride again takes it from the ether; so that the section is thus gradually dried, and with as little shrinking as possible, however delicate it may be. This process is most fitted for the preparation of succulent roots, tubers, or stems, and indeed is only fit for those tissues which are not injured by immersion in ether, or dissolved by it, such as fat, &c., or colouring matter.

Dissection.—As I stated at the commencement of this chapter, no written instructions can enable any student to become an adept in this branch without much experience and no little study. I will, however, describe the necessary apparatus, and afterwards mention the mode of treatment which certain objects require.

A different microscope is manufactured for the purpose of dissection, most first-rate makers having their own model. The object-glasses of many of these are simple, and consequently not expensive; but one of the great requisites is a stage large enough to hold the trough, in which the operation is often performed. Where this is the case it would scarcely be worth the expense of getting a dissecting microscope if the student were pursuing no particular study, but merely used the instrument when an object to be operated upon turned up accidentally. The ordinary form is much improved for this purpose, by having two wooden rests placed at the sides of the microscope, upon which the hands may be supported when working upon the stage. They should be weighty enough to be free from danger of moving. These supports will also be found to remedy much of the weariness which inevitably arises from having to sustain the hands as well as work with them. The erector, as I before observed, is necessary to a young student; but with a little practice he may work very well without it.

We will now notice some of the instruments which are most useful in dissection. Two or three different sizes of ordinary scissors should be possessed, but the shapes must be as modified in others for many purposes, as those used by surgeons; a pair with the cutting parts bent in a horizontal direction, and another pair slightly curved in a perpendicular; so that parts of the substance operated upon may be reached, which it would be impossible to touch with straight scissors. One point of these is sometimes blunt, and the other acute, being thus made very useful in opening tubular formations. Another form of these is made, where the blades of the scissors are kept open by a spring, the handles being pressed together by the fingers. Where it is desirable, one or both of these handles may be lengthened to any degree by the addition of small pieces of wood.

THE KNIVES which are most useful are those of the smallest kind which surgeons employ in very delicate operations. These are made about the length of an ordinary pen-knife, and are fixed in rather long flattish handles; some are curved inwards, like the blade of a scythe, others backwards; some taper to a point, whilst others again are broad and very much rounded. Complete boxes are now fitted up by the cutlers, of excellent quality and surprisingly cheap.

NEEDLES.—These are very useful and should be firmly fixed in handles as recommended in Chapter II. It is convenient to have them of various lengths and thicknesses. If curved by heating and bending to any required shape they may be re-hardened by putting them whilst hot into cold water. Dr. Carpenter also makes edged instruments by rubbing down needles upon a hone. They are more pleasant to work with when *short*, as the spring they have whilst *long* robs them of much of their firmness. Glass points made by drawing out glass rods to a point will be found useful in manipulating with acids.

A *glass syringe* is also useful in many operations, serving not only to cleanse the objects but to add to, or withdraw liquids from, the *dissecting-trough*. This trough will now be described, as many substances are so changed by becoming dry that it is impossible to dissect them unless they are immersed in water during the operation. If the object be opaque and must be worked by reflected light, a small square trough may be made to the required size, of gutta-percha, which substance will not injure the edge of the knives, &c.; but where transparency is necessary, a piece of thin plate-glass must be taken, and by the aid of marine-glue (as explained in Chapter V.) sides affixed of the required depth. As pins, &c., cannot be used with the glass troughs and the substance must be kept extended, a thin sheet of cork loaded with lead in order to keep it under water may be used; but this, of course, renders the bottom opaque. When working with many thin substances, a plate of glass three or four inches long and two wide will serve every purpose, and be more pleasant to use than the trough. A drop or two of water will be as much liquid as is needed, and this will lie very well upon the flat surface. As these are the principal apparatus and arrangements which are requisite in dissection, the method of proceeding in a few cases may now be noticed.

Vegetables.—The dissection of vegetable matter is much less complicated than that of animal; maceration in water being a great assistant, and in many cases removing all necessity for the use of the knife, especially if hot water can be used without injury to the objects, as is the case with many. This maceration may be assisted by needles, and portions of the matter which are not required may be removed by them. When, for instance, the spiral vessels which are found in rhubarb are wanted, some parts containing these are chosen and left in a small quantity of water until the mass becomes soft, and this is more quickly effected when the water is not changed. The mass must be then placed upon a glass plate when practicable, or in the trough when large, and with the aid of two needles the matter may be removed from the spiral vessels, which are plainly seen with a comparatively low power; and by conveying these to a clean slip of glass, repeating the process, and at last washing well, good specimens may be procured. Most of these should be mounted in some of the preservative liquids in the manner described in Chapter V. Many, however, may be dried on the slide, immersed in turpentine, and then mounted in balsam; but liquid is preferable, as it best preserves their natural appearance. Certain kinds of vegetables require a different treatment to separate these spiral vessels. Asparagus is composed of very hard vegetable matter, and some have recommended the stems to be first boiled, which will soften them to such a degree that they may easily be separated. Dilute acids are also occasionally used to effect this; and in some instances to obtain the *raphides* caustic potash may be employed; but after *any* of these agents have been used, the objects must be thoroughly cleansed with water, else the dissecting instruments (and perhaps the cell) will be injured by the action of the remaining portion of the softening agent.

For the dissection of *animal tissues* it is necessary that the instruments be in the best order as to sharpness, &c.; and as the rules to be observed must necessarily be somewhat alike in many instances, the treatment required by some of the objects most frequently mounted will now be described. We may here remark that *cartilage* can be best examined by taking sections which will show the arrangement of the cells very perfectly. This, however, is plainly seen in the mouse's ear without any section being necessary. Glycerine, the preservative liquids before mentioned, and Canada balsam are all used to mount it but perhaps the first-named may be preferred in many cases.

Before treating of separate objects it will be well to notice what M. Brunetti has said on preparing *anatomical specimens.* The process consists of four stages—viz., washing, divesting of fat, treating with tannin, and desiccation. A stream of pure water is injected through the blood-vessels and secretory ducts of the part to be preserved; the water is afterwards expelled by means of alcohol. To remove the fat, the vessels are in like manner injected with ether, which penetrates the tissues and dissolves all the fatty matters. These operations occupy about two hours, and the object thus prepared may then be kept for a long time in ether, if desired. A solution of tannin is next injected in a similar manner, and the ether washed out by a stream of pure water. The preparation is then placed in a double-bottomed vessel containing boiling water—a sort of *bain-marie*—in order to displace the fluid previously used by dry heated air. Air compressed in a reservoir to about two atmospheres is forced into the vessels and ducts through heated tubes containing chloride of calcium: all moisture is thus expelled and the process is completed. The preparation thus treated is light, and retains its volume, its normal consistence, and all its histological elements.

MUSCLE.—This is what is commonly called the flesh of animals. If a piece be laid upon the slide under the microscope, bundles of fibres will be perceived, which with needles and a little patience may be separated into portions, some of these being striated, or marked with alternate spaces of dark and light. Some of the *non-striated* or *smooth* class of muscle, such as is found in intestines, may be prepared for the microscope by immersing for a day or two in nitric acid diluted with three or four parts of water, and then separating with needles and mounting as soon as possible. Sometimes *boiling* is resorted to to facilitate the separation, and occasions little or no alteration in the material.

Specimens are often taken from the *frog* and the *pig*, as being amongst the best, *Goadby's Solution* being generally used in mounting them. The muscles of insects also show the striæ very perfectly.

Nerve-tissue.—This is seldom mounted; as Dr. Carpenter observes, "no method of preserving the nerve-tissue has been devised which makes it worth while to mount preparations for the sake of displaying its minute characters," but we will mention a few particulars to be observed in its treatment. The nerve should be taken from the animal as soon as possible after death, and laid upon a glass slide, with a drop or two of serum if possible. The needles may be used to clean it, but extreme delicacy is necessary. It will be found that the nerve is tubular and filled with a substance which is readily ejected by very slight pressure. When the nerve is submitted to the action of acetic acid, the outer covering, which is very thin, is considerably contracted, whilst the inner tube is left projecting; and thus is most distinctly shown the nature of the arrangement. Dr. Lockhart Clarke, who has made great researches into the structure of the spinal cord, gives a part of his experience as follows:—He takes a perfectly fresh spinal cord and submits it to the action of strong spirits of wine. This gives the substance such a degree of hardness that thin sections may be readily cut from it, which should be placed upon a glass in a liquid consisting of three parts of spirit and one of acetic acid, which renders them very distinct. M. Grandry has treated *nerve-tissue* thus:—Taking portions of nervous tissue obtained from the frog and rabbit, he placed them in a one-fourth per cent. solution of nitrate of silver in pure water, macerating them for five days in the dark, and then exposing them for three days to bright light. If the surface of the cord thus treated be carefully teazed out with needles, the axis-cylinders are found to exhibit a very regular and sharply defined transverse striation—clear, unstained striæ alternating with deeply tinted ones. Dr. Bastian recommends us to mount delicate specimens of nerve-tissue in a mixture of glycerine and carbolic acid in the proportion of fifteen of the first to one of the second. To mount these sections, they must now be steeped in pure spirit for two hours, and afterwards in oil of turpentine, and lastly must be mounted in Canada balsam.

Dr. Lionel Beale recommends the use of chloride of gold for colouring nerve fibres. A solution containing from 2 to 1 per cent. in distilled water should be made. The tissue having been soaked in it until it becomes straw-coloured, is to be washed, and then placed in very dilute acetic acid containing one per cent. or less. The nerves exhibit a blue or violet tinge on exposure to light for a few hours. He speaks also of solution of osmic acid for the same purpose, 1 part to 100 of water, but not with much approval. The aniline colours, such as magenta and solferino, may, according to the same authority, be also employed for most tissues. They are not very soluble in water, but are readily dissolved by alcohol. A grain of the colour, 10 to 15 minims of alcohol, and an ounce of distilled water, make a dark red or blue (purple) solution which colours tissues very readily. For these and many other useful formulæ for the same purpose, the reader may consult "Beale on the Microscope."

Dr. Klein, in No. 40 of the *Monthly Microscopical Journal*, in order to demonstrate the nerves of the cornea, takes that of the rabbit or guinea-pig, a quarter or half an hour after death, and places it in a half per cent. solution of chloride of gold, for from one and a half to two hours—that of the guinea-pig for an hour to an hour and a quarter. After that, the cornea is washed in distilled water, and exposed to the light in distilled water for from 24 to 36 hours (the water being changed twice, or oftener). After this time has elapsed, the cornea is transferred into a mixture of one part pure glycerine and two of distilled water, where it remains for two or three days. Up to this time the cornea has not assumed a darker colour than ash-grey, perhaps having a violet tint; at all events the whole of the cornea is transparent. It is then brushed over on its anterior surface under water with a fine camel-hair brush very gently, so as to remove the precipitates of the gold salt. Sections of a cornea so prepared may be made on the finger by a sharp razor, and must be examined and kept in glycerine.

LIVER, KIDNEY, SPLEEN, LUNG, &c.—Some parts which are too soft to be cut into sections in their ordinary state, are usually hardened by being steeped in a solution of chromic acid, about two grains to an ounce of water. This will take some weeks according to the substance, and the solution should be changed now and then. Dr. Bastian, for mounting, uses Canada balsam partially dried to dispel the turpentine, and then diluted to necessary consistence with benzole. The section being cut from the hardened organ is washed in spirits of wine for some minutes, then a drop of liquid carbolic acid is placed on the slide where the specimen is to be mounted. Take the specimen and let its edge touch a piece of blotting-paper, and place it upon the carbolic acid, which will render a thin section transparent in about a minute. Remove the superfluous acid with blotting-paper, when two or three drops of chloroform must be poured upon the section and remain one minute. Drain off and place upon the object the solution of Canada balsam in benzole, and apply the thin glass cover. Or place the object in ordinary spirits of wine for about a minute to wash it, then remove into absolute alcohol for five minutes. Lay it upon the slide and drain, cover with one or two drops of benzole for about a minute, tilt to drain off, and proceed as above.

Both these methods are good, but the first does not always answer for sections of liver, as they generally are acted upon by carbolic acid; but few other tissues are thus affected. Tinted specimens seem equally safe when mounted in this way.

SECTIONS OF BRAIN AND SPINAL CORD.—Dr. Bastian gives his experience of these tissues as follows:—I immerse the section for about ten minutes in absolute alcohol diluted with eight per cent. of water, then place upon the glass slide, and before it becomes dry pour over it two or three drops of pyroacetic acid for about half a minute. Tilt this off and replace by chloroform. Watch the effects, as before, under the microscope, and then cover with the Canada balsam solution and finish. These specimens, however, are not always permanent in their appearance, according to the results of some.

Mr. Alfred Sanders gives his experience as differing somewhat from this. He says—The brain, or other structure, being, as usual, hardened in chromic acid, the section is put for a short time in spirits of wine, and thence transferred to the creosote, which makes it transparent in a few minutes, when it is placed in Canada balsam. The balsam will mix easily with the creosote, or the solution in benzole may be employed.

Tracheæ of Insects, &c.—The nature of these was described in Chapter IV., but the method of procuring them was not explained, as this clearly belongs to dissection. The larger tubes are readily separated by placing the insect in water, and fixing as firmly as possible, when the body must be opened and the viscera removed. The tracheæ may then be cleaned by the aid of a camel-hair pencil, and floated upon a glass, where they must first be allowed to dry, and then be mounted in balsam. Mr. Quekett gives the following method of removing the tracheæ from the larva of an insect:—"Make a small opening in its body, and then place it in strong acetic acid. This will soften or decompose all the viscera, and the tracheæ may then be well washed with the syringe, and removed from the body with the greatest facility, by cutting away the connections of the main tubes with the spiracles by means of fine-pointed scissors. In order to get them upon the slide, it must be put into the fluid, and the tracheæ floated upon it; after which they may be laid out in their proper position, then dried and mounted in balsam." If we wish them to bear their *natural* appearance, they must be mounted in a cell with Goadby's fluid; but the structure is *sometimes* well shown in specimens mounted dry. As before mentioned, these tracheæ terminate on the outside in openings termed spiracles, which are round, oblong, and of various shapes. Over these are generally a quantity of minute hairs, forming a guard against the entrance of dust. The forms of these are seldom alike in two different kinds of insects, so that there is here a wide field for the student. The dissection, moreover, is very easy, as they may be cut from the body with a sharp knife or scissors, and mounted in balsam or fluid. Many of the larvæ afford good specimens, as do also some of the common Coleopterous insects. Perhaps, no more satisfactory object can be met with to give the student good examples of *spiracles* than the water-beetle Dytiscus, before mentioned, as affording such perfectly beautiful suckers. They will be found to vary in appearance according to the part of the body from which they are taken; but all are equally interesting.

Mr. Lewis G. Mills, LL.B., gives the following account of his extracting the poison glands from a spider:—Having killed a large spider with chloroform, I left it in water for seven or eight days. This treatment usually softens the outer skin of insects and causes the viscera to swell, so as to burst through the outer integument, and it is in this state, perhaps, that the *poison glands* are most easily discovered and traced to their points of attachment. I then drew the mandibles from the body, and, having placed them with a little water on a slide and covered them with a piece of thin glass, I found that, upon the application of pressure, the two glands shot out and protruded from the bases of the mandibles. I tore open one of the mandibles with needles, so as to disturb the gland as little as possible. The gland then appeared as a closed sac, attached by a hollow cord, about the length of the gland itself, to the base of the fang, where also was a large bundle of muscular fibre.

Fish.—The most interesting part of fish to the microscopic anatomist is undoubtedly the *breathing* apparatus.

It is not a very difficult matter to open the head and remove the gills, which are very beautiful. Under the outer covers lie a quantity of thin plates or leaves (as of a book) which in different fishes are of various shapes, but are made like net-work by the numerous veins and arteries which convey the blood to be acted upon by the air and gases in the water, as is done in the lungs of a man. These plates are of such numbers that in a good-sized salmon the surface exposed has been estimated at two thousand square inches, *i.e.*, about fourteen square feet. The beauty of these is, of course, not *perfectly* shown until they are *injected*, which will be noticed elsewhere.

Tongues, or Palates, of Molluscs.—Of the nature of these, Dr. Carpenter gives the following description:—"The organ which is commonly known under this designation is one of a very singular nature; and we should be altogether wrong in conceiving of it as having any likeness to that on which our ordinary ideas of such an organ are founded. For, instead of being a projecting body, lying in the cavity of the mouth, it is a tube that passes backwards and downwards beneath the mouth, its higher end being closed, whilst in front it opens obliquely upon the floor of the mouth, being, as it were, slit up and spread out so as to form a nearly flat surface. On the interior of the tube, as well as on the flat expansion of it, we find numerous transverse rows of minute teeth, which are set upon flattened plates; each principal tooth sometimes having a basal plate of its own, whilst in other instances one plate carries several teeth." These palates, or tongues, differ much amongst the Gasteropods in form and size, some of them being comparatively of an immense length. Many are amongst the most beautiful objects when examined with polarized light. They must, however, be procured by dissection, which is usually performed as follows:—The animal is placed on the cork in the dissecting-trough before mentioned, and the head and forepart cut open, spread out, and firmly pinned down. With the aid of fine scissors or knife, the tongue must be then detached from its fastenings, and placed in water for a day or two, when all foreign matter may with a little care be removed. In what way it should be mounted will depend on the purpose for which it is intended. If for examination as an ordinary object, it may be laid upon the slide and allowed to dry, which arrangement will show the teeth very well. If we wish to see it as it is naturally, it must be mounted in a cell with Goadby's fluid; but if it is wanted as a polarizing object, it must be floated upon a slide, allowed to dry thoroughly, and then Canada balsam added in the usual manner.

In the stomach, also, of some of these molluscs teeth are found, which are very interesting objects to examine, and must be dissected out in the same manner as the tongues.

Since writing the above, Dr. Alcock (whose very beautiful specimens prove him to be a great authority in this branch) has published some of his experience in the second volume of the third series of "Memoirs of the Literary and Philosophical Society of Manchester." By his permission I make the following extract:—

"This closes my present communication on the tongues of mollusca; but as some members may possibly feel inclined to enter upon the inquiry themselves, I think it will not be amiss to add a few remarks on the manner in which they are to be obtained.

"First, as to the kinds best worth the trouble of preparation. Whelks, Limpets, and Trochuses should be taken first. Land and fresh-water snails can scarcely be recommended, except as a special study,—their tongues being rather more difficult to find, and the teeth so small that they require a high power to show them properly. It would appear, from Spallanzani's description of the anatomy of the head of the snail, that even he did not make out this part, although, in his curious observations on the reproduction of lost parts, he must have carefully dissected more snails than any other man.

"As to preserving the animals till wanted, they should simply be dropped alive into glycerine or alcohol. Glycerine is perhaps best where only the tongues are wanted; but it leaves the animals very soft; and as it does not harden their mucus at all, they are very slippery and difficult to work upon when so preserved.

"Then as to the apparatus required for dissection. In the first place, all the work is to be done under water, and a common saucer is generally the most convenient vessel to use. No kind of fastening down or pinning out of the animal is needed; and, in fact, it is much better to have it quite free, that you may turn it about any way you wish. The necessary instruments are a needle-point, a pair of fine-pointed scissors, and small forceps; the forceps should have their points slightly turned in towards each other.

"A word or two on the lingual apparatus generally, and on its special characters in a few different animals, will conclude what I have to say.

"The mode of using the tongue can be easily seen in any of the common water-snails, when they are crawling on the glass sides of an aquarium; it may then be observed that from between the fleshy lips a thick mass is protruded, with a motion forwards and upwards, and afterwards withdrawn, these movements being almost continually repeated. The action has the appearance of licking; but when the light falls suitably on the protruded structure, it is seen to be armed with a number of bright points, which are the lingual teeth, so arranged as to give the organ the character and action of a rasp.

"If you proceed to dissection, and open the head of one of these mollusca (say, for instance, a common limpet), you will find the cavity of the mouth almost filled with the thick fleshy mass, the front of which is protruded in the act of feeding; and on its upper surface, extending along the middle line from back to front, is seen the strong membranous band upon which the teeth are set. The mass itself consists of a cartilaginous frame, surrounded by strong muscles; and these structures constitute the whole of the active part of the lingual apparatus.

"But the peculiarity of the toothed membrane, which makes its name of *ribbon* so appropriate, is, that there is always a considerable length of it behind the mouth, perfectly formed, and ready to come forward and supply the place of that at the front, which is continually wearing away by use.

"In the limpet this reserve ribbon is of great length, being nearly twice as long as the body, and the whole of it is exposed to view on simply removing the foot of the animal; nothing, then, can be easier than to extract the tongue of the common limpet. But, unfortunately, what you find in one kind of mollusc is not at all what you find in another. In the Acmæas, for instance, which are very closely related to the limpets, and have shells which cannot be distinguished, the reserve portion of the ribbon has to be dug out from the substance of the liver, in which it is imbedded, that organ being, as it were, stitched completely through by a long loop of it. . . . . . . It might be thought a comfortable reflection that, at all events, one end of the ribbon can always be found in the mouth; but in many cases this is about the worst place to look for it. Perhaps it may appear strange that in some of the smaller species, with a retractile trunk, a beginner may very likely fail altogether in his attempt to find the mouth; if, however, the skin of the back be removed, commencing just behind the tentacles, there will be very little difficulty in making out the trunk, which either contains the whole of the ribbon, as in the whelk, or the front part of it, as in *Purpura* and *Murex*, where a free coil is also seen to hang from its hinder extremity. . . . . . . In the periwinkles the same plan of proceeding, by at once opening the back of the animal, is best: and on doing so, the long ribbon, coiled up like a watch-spring, cannot fail to be found.

"In the Trochuses, and indeed in all the *Scutibranchiata*, one point of the scissors should be introduced into the mouth of the animal, and an incision made directly backwards in the middle line above to some distance behind the tentacles; the tongue is then immediately brought into view, lying along the floor of the mouth."

Dr. Alcock's method of dissection will be found to differ in some degree from the general rules before given; and when the tongue is dissected out he washes it for one hour (shaking it now and then) in a weak solution of potash. After cleaning thoroughly in water, it must be mounted by one of the methods before mentioned.

Mr. Edwards, of New York, no mean authority, gives his experience as follows:—I use a rather strong solution of caustic potassa, the strength of which I cannot specify as it must differ with the species under manipulation, as some ribbons (or tongues) are injured much sooner than others. Plunge the whole animal in this solution; in the case of very small creatures shell and all. I have found it better to let the animal stand until it dies and begins to decompose, when it can readily be removed and falls in pieces. The lingual ribbon is not so easily decomposed. Now place and leave the animal in the potassa solution for some days, or boil at once. Almost everything is now dissolved but the shell, some few fragments, and the desired ribbon. Wash carefully with fresh water, and if it is to be preserved before mounting, remove to alcohol. To mount it, remove from the spirit and boil a short time in turpentine, when it can be put up in Canada balsam. Mr. May expresses himself as "*standing utterly aghast*" at any man so interfering with nature as to put up these objects in balsam, thus pressing and destroying their true forms. He recommends a cell and a weak form of Goadby's solution.

Amongst insects, especially the grasshopper tribe, are found many which possess a gizzard, armed with strong teeth, somewhat similar to those of the molluscs. It requires great nicety of manipulation to obtain these for the microscope; but Mr. L. G. Mills, before quoted, gives the following instructions:—Kill the insect with chloroform and place it in a vessel of water. Hold it down firmly with a pair of tweezers, and with the back of a dissecting knife draw the head steadily from the body. The head brings with it the stomach, gizzard, and chief portion of the digestive tubes. Place all these under a dissecting microscope, when the gizzard, being just below the stomach and darker in colour, is easily distinguished, and may be separated by two cuts with the knife. It then forms a short tube, the teeth being inside. The opening-out of this tube, especially if it be small, requires delicate handling: if the point of a fine knife can be fairly inserted, then one firm cut downward upon the glass will lay open the gizzard. Here great care is needed; and sometimes it is well to put a fine needle up the tube, and cut down upon the needle. Among the small weevils the membrane is delicate, so that great care is necessary.

We have now considered most of those objects which require any *peculiar* treatment in section-cutting, &c.; but in no branch of microscopic manipulation is experience more necessary than in this.

CHAPTER VII.

INJECTION.

1. INJECTION is the filling of the arteries, veins, or other vessels of animals with some coloured substance, in order that their natural arrangement may be made visible. This is, of course, a delicate operation, and needs special apparatus, which I will now attempt to describe.

2. *Syringe.*—This is usually made to contain about two ounces. On each side of the part next to the handle is a ring, so that a finger may be thrust through it, and the thumb may work the piston as in an ordinary syringe. The plug of the piston must be packed with soft leather well oiled or greased, in order to free it from all danger of any liquid penetrating it, and fit so closely as to be perfectly air-tight; and if, when it has been used awhile, it is found that some of the liquid escapes past the plug into the back part of the body, it must be *repacked,* which operation will be best understood by examining the part. These syringes are made of various sizes, but in ordinary operations the above will be all that is needed. The *nozzle* is about an inch long, and polished so accurately that there is no escape when the *pipes* are tightly placed upon it *dry.*

3. The *pipes* are usually about an inch long, to their ends are affixed thicker tubes so as to fit the nozzle, as before mentioned, whilst a short arm projects from each side of these, so that the silk or thread which is used to tie the artery upon the thin pipe, may be carried round these arms, and all danger of slipping off prevented. The *pipes* are made of different sizes, from that which will admit only of a very fine needle (and this will need now and then to be cleaned, or to be freed from any chance obstruction), to

that which will take a large pin. These sizes must always be at hand, as the vessels of some subjects are exceedingly minute.

4. *Stopcock.*—This is a short pipe like a small *straight* tap, which fits accurately upon the end of the syringe like the pipes, and also takes the pipes in the same manner. The use of this is absolutely necessary when the object is so large that one syringe full of liquid will not fill it. If no preventive were used, some part of the liquid would return whilst the syringe was being replenished, but the stopcock is then turned as in an ordinary tap, and all danger of this effectually removed.

5. *Curved needles.*—These are easily made by heating common needles at the end where the eye is situated, and bending them with a small pair of pliers into a segment of a circle half an inch in diameter. They are, perhaps, more convenient when the bent part is thrown slightly back where it commences. The pointed end is then thrust into a common penholder, and the needle needs no re-tempering, as the work for which it is wanted is simply to convey the thread or silk *under* any artery or vessel where it would be impossible to reach with unassisted fingers.

6. A kind of *forceps*, commonly known by the name of "bulldog forceps," will be constantly required during the process of injecting. These are short, usually very strong, but not heavy, and close very tightly by their own spring, which may be easily overcome and the artery so released by the pressure of the fingers. When any vessel has not been tied by the operator, and he finds the injected fluid escaping, one of these "bulldogs" may be taken up and allowed to close upon the opening. This will cause very little interruption, and the stoppage will be almost as effectual as if it were tied.

7. When the ordinary mode of injection is employed, it is necessary that the preparations be kept warm during the time they are used, otherwise the gelatine or size which they contain becomes stiff, and will not allow of being worked by

the syringe. For this purpose we must procure small earthenware or tin pots of the size required, which will differ according to the kind of work to be done; and to each of these a loose lid should be adapted to protect it from dust, &c. These pots must be allowed to stand in a tin bath of water, under which a lamp or gas flame may be placed to keep the temperature sufficiently high to insure the perfect fluidity of tbe mixture. The tin bath is, perhaps, most convenient when made like a small shallow cistern; but some close it on the top to place the pots upon it, and alter the shape to their own convenience.

8. We will now inquire into some of the materials which are needed in this operation; the first of which is *size*. This substance is often used in the form of *glue*, but it must be of the very best and most transparent kind. To make the liquid which is to receive the colours for the usual mode of injecting, take of this glue seven ounces, and pour upon it one quart of clean water; allow this to stand a few hours, and then boil gently until it is thoroughly dissolved, stirring with a wooden or glass rod during the process. Take all impurities from the surface, and strain through flannel or other fine medium. The weather affects this a little as to its stiffness when cold, but this must be counteracted by adding a little more glue if found too liquid.

9. Instead of glue, gelatine is generally used, especially when the work to be accomplished is of the finer kind. The proportions are very different in this case, one ounce of gelatine to about fourteen ounces of water being sufficient. This, like glue, must be soaked a few hours in a small part of the cold water, the remainder being boiled and added, when it must be stirred until dissolved. A good size may be made by boiling clean strips of parchment for awhile, and then straining the liquid whilst hot through flannel; but when the injections are to be *transparent*, it is of the greatest importance that the size be as colourless as possible. For this purpose good gelatine must be employed,

as Nelson's or Cox's: some persons of experience prefer the latter.

10. *Colours.*—The size-solution above mentioned will need some colouring matter to render it visible when injected into the vessels of an animal, and different colours are used when two or more kinds of vessels are so treated, in order that each set may be easily distinguished by sight. The proportion in which these colours are added to the size-solution may be given as follows:—

11. For

Red	8	parts of size-solution (by weight)	to	1	part of	vermilion.
Yellow...	6	„	„	1	„	chrome yellow.
White ...	5	„	„	1	„	flake-white.
Blue	3	„	„	1	„	blue-smalt, fine.
Black ...	12	„	„	1	„	lampblack.

Whichever of these colours is used must be levigated in a mortar with the addition of a very small quantity of water until every lump of colour or foreign matter is reduced to the finest state possible, otherwise in the process of injecting it will most likely be found that some of the small channels have been closed and the progress of the liquid stopped. When this fineness of particles is attained, warmth sufficient to render the size quite fluid must be used, and the colour added gradually, stirring all the time with a rod. It may be here mentioned that where one colour only is required, vermilion is, perhaps, the best; and blue is seldom used for opaque objects, as it reflects very little more light than black.

12. When it is wished to fill the capillaries (the minute vessels connecting the arteries with the veins), the Micrographic Dictionary recommends the colouring matter to be made by double decomposition. As a professed handbook would be, perhaps, deemed incomplete without some directions as to the mode of getting these colours, I will here

use those given in that work. For red, however, vermilion, as above stated, may be used; but it must be carefully examined by reflected light to see whether it be free from all colourless crystals or not. It must first be worked in a mortar, and then the whole thrown into a quantity of water and stirred about; after leaving it not longer than a quarter of a minute, the larger portions will settle to the bottom, and the liquid being poured off will contain the finer powder. This may then be dried slowly, or added to the size whilst wet in the manner before advised.

13. *Yellow injection.*—To prepare this, take—

Acetate (sugar) of lead	380 grains.
Bichromate of potash	152 ,,
Size	8 ounces.

Dissolve the lead salt in the warm size, then add the bichromate of potash finely powdered.

Some of the chromic acid remains free, and is wasted in this solution, so the following is given :—

Acetate of lead.....................	190 grains.
Chromate of potash (neutral) ...	100 ,,
Size	4 ounces.

The first of these has the deepest colour, and is the most generally used.

14. *White injection.*—This is a carbonate of lead :—

Acetate of lead	190 grains.
Carbonate of potash...............	83 ,,
Size	4 ounces.

Dissolve the acetate of lead in the warm size, and filter through flannel; dissolve the carbonate of potash in the smallest quantity of water, and add to the size: 143 grains of carbonate of soda may be substituted for the carbonate of potash.

15. For blue injection, which is not, however, much used with reflected light, as before stated, take—

Prussian blue........................	73 grains.
Oxalic acid...........................	73 ,,
Size	4 ounces.

The oxalic acid is first finely powdered in a mortar, the Prussian blue and a little water added, and the whole then thoroughly mixed with the size.

16. It may here be repeated, that it is only when the capillaries are to be filled that there is any need to be at the trouble to prepare the colours by this double decomposition; and, indeed, colours ground so finely may be procured that the above instructions would have been omitted, had it not been supposed that some students might find a double pleasure in performing as much of the work as possible by their own unaided labours.

17. The process of injection may now be considered; but it is impossible for written instructions to supply the place of experience. I will do my best, however, to set the novice at least in the right way. There are two kinds of injection—one where the object and colours are opaque, and consequently fit for examination by *reflected* light only; the other, where the vessels are filled with transparent colours, and must be viewed by *transmitted* light. The first of these is most frequently employed, so we will begin with it. In the object which is to be injected, a vessel of the kind which we wish to be filled must be found; an opening must then be made in it to allow one of the small pipes before mentioned to be thrust some distance within it. When this is accomplished, thread the curved needle with a piece of silk thread, or very fine string, which some operators rub well with beeswax. This thread must not be too thin, else there is danger of cutting the vessel. The cord is then carried under the inserted pipe, and the vessel bound tightly upon it, the ends being brought up round the transverse arms,

and there tied; so that all danger of accidentally withdrawing the pipe is obviated. Care must now be used in closing all the vessels which communicate with that where the pipe is placed lest the injecting fluid escape; and this must be done by tying them with silk. Should, however, any of these be left open by accident, the bulldog forceps must be used, as before recommended.

18. The part to be injected must now be immersed in warm water, not, however, above 100° Fahrenheit, and left until the whole is thoroughly warmed. Whilst this is being done, the coloured size must be made ready by the pot being placed in the tin bath of warm water, which must be of sufficient temperature (about 110° Fahrenheit) to keep it perfectly liquid. For the same purpose, the syringe is often tightly covered with two or three folds of flannel; and, indeed, there is no part of the process which requires more attention. If the substance to be injected is too hot, it is injured; whilst, if any of the articles are too cold, the gelatine, or size, loses a part of its fluidity, and consequently cannot enter the minute vessels. When all is prepared, the syringe, with the stopcock attached, should be warmed, and then filled and emptied with the injecting fluid two or three times, care being taken that the end of the syringe be kept beneath any bubbles which form upon the surface. The syringe may then be filled, and closely attached to the pipe which is tied in the vessel. With a firm and steady pressure the piston must be forced downwards, when the substance will be perceived to swell, and the colour show itself in places where the covering is thin. When the syringe is *almost* emptied of its contents, the stopcock must be turned to prevent any escape of the injection from the subject. It must then be refilled, as in the first instance, and the process repeated. I say *almost emptied*, because it is well not to force the piston of the syringe quite to the bottom, lest the small quantity of air which frequently remains be driven into some of the vessels, and the object be injured or quite ruined. As the injection is continued, it will be found that

the force required grows greater, yet care must be taken not to use too much, or the vessels will burst, and render all the labour fruitless. The movement of the piston must be occasionally so slow as to be almost imperceptible, and for this reason the piston-rod is sometimes marked with lines about one-eighth of an inch apart.

19. Of course, during the whole process the injecting fluid and subject must be kept at a temperature high enough to allow the liquid to flow freely; and the escape of a little of it need cause no fears to the operator, as it is almost impossible to fill a subject without some loss. When the injected object has received sufficient fluid, it should have a plump appearance, owing to all the vessels being well filled. The vessel must then be tied up where the pipe was inserted, and the whole left in cold water two or three hours, after which time it may be mounted; but it may be well to notice a few things which the beginner ought to know before entering into that part of the process; and he may be here informed that it is not necessary to mount the objects immediately, otherwise it would be impossible for one person to make use of half of any large subject, as it would be in a state of decay long before each part could have been examined and separated. Large pieces should be therefore immersed in equal parts of spirits of wine and water, or glycerine, which some think better still, and thus preserved in bottles until time can be given to a closer examination.

20. In operating upon large subjects, entire animals, &c., the constant pressure required by the piston of the syringe grows wearisome, besides occupying both hands, which is sometimes inconvenient when working without assistance. To obviate this, another way of driving the syringe was published in the Micrographic Dictionary which I will quote here:—"We have therefore contrived a very simple piece of apparatus, which any one can prepare for himself, and which effects the object by mechanical means. It consists of a rectangular piece of board, two feet long and ten inches wide, to one end of which is fastened an inclined

piece of wood (equal in width to the long board, and one foot high). The inclined portion is pierced with three holes, one above the other, into either of which the syringe may be placed—the uppermost being used for the larger, the lowermost for the smaller syringe; and these holes are of such size as freely to admit the syringe covered with flannel, but not to allow the rings to pass through them. The lower part of the syringe is supported upon a semiannular piece of wood, fastened to the upper end of an upright rod, which slides in a hollow cylinder fixed at its base to a small rectangular piece of wood; and by means of a horizontal wooden screw, the rod may be made to support the syringe at any height required. The handle of the syringe is let into a groove in a stout wooden rod connected by means of two catgut strings with a smaller rod, to the middle of which is fastened a string playing over a pulley, and at the end of which is a hook for supporting weights, the catgut strings passing through a longitudinal slit in the inclined piece of wood." When in use the syringe is filled with injecting fluid, and passed through one of the three holes which is most suitable. The object being placed so that the pipe and syringe can be best joined, the rod and strings are set in order, and a weight placed on the hook. The stop-cock must then be opened gradually, when the operator will be able to judge whether the weight is a proper one or not: if the piston is driven with any speed, there is danger of injuring the subject, and less weight may be used; if, however, the piston do not move, more must be added.*

21. Such is the method recommended in the Micrographic Dictionary, and perhaps it is as good as any mechanical plan could be; but where the operator is willing to undergo the labour of performing all this with the hand, he has a much better chance of succeeding, because the pressure can be regulated so accurately, and changed so

* There is in the Monthly Microscopic Journal, vol. ii., page 48, another ingenious apparatus for injecting purposes.

quickly when requisite, that no mere machine can compete with it, however well contrived.

22. When the beginner attempts to inject a subject, one of his difficulties is finding the vessel from which to commence. Another consists in distinguishing the arteries from the veins; but this is partly removed by making a longitudinal incision in the vessel, and with a blunt thick needle probing a little distance into the tube. The artery will be found thicker in the coating than the vein, and the difference is easily perceived by this mode of testing: the vein is also of a bluer colour than the artery. I say above, a "longitudinal incision" must be made: the reason for this is, that an artery when cut across contracts considerably, and is lost in the adjoining substance; but where the opening is made *longitudinally* all danger of this contraction is obviated.

23. The different systems of vessels are often injected with various colours, so that their relative positions, may be shown more clearly. In some specimens, the veins are injected with white, and the arteries with red; in the kidney, the urinary tubes are often filled with white, and the arteries with red. Then, again, the liver affords tubes for three or four colours. But no written instructions on this point can benefit the young student, and he must be content for a while to employ himself with single colours until he has gained the mechanical skill and the primary knowledge which are necessary before he can make any advance.

24. We will now consider the best methods of *mounting* injected objects. They must always be well washed in water after they have been kept in any preservative liquid, using a camel-hair pencil to clean the surface if necessary. Many parts when injected are in masses, such as the lungs, liver, &c., of animals, and consequently sections of these must be cut. For this purpose Valentin's knife is very convenient, as the thickness can be regulated so easily; but where the injections are opaque, there is no need to have the

sections very thin. Some few of this kind undergo comparatively little change in drying, so that the section may be well washed and floated upon the glass slide in the place desired, where it will dry perfectly and adhere to it. It must then be moistened with turpentine and mounted in Canada balsam like other objects. No great heat should be used with these preparations, as it is very liable to injure them; and some of the colours seem to suffer a slight contraction when any great degree of warmth is applied. There are many objects, however, which must be seen in the mass to be understood, and, indeed, lose all their form and beauty in drying, such as certain parts of the intestines, &c. These must be mounted in fluid, with the precautions noticed at length in Chapter V., and for this purpose either Goadby's fluid, the chloride of zinc solution, or spirit diluted with ten parts of distilled water, may be employed. It is a good thing, when practicable, to mount similar objects on two separate slides, using different preservative liquids, and taking the precaution of marking each with the kind of liquid employed. This not only serves as a guide to what is best for certain subjects, but if one is injured, there will probably be a good specimen in the other.

25. It may be here mentioned that many are now mounting sections of injected substances with the balsam and chloroform before mentioned, instead of using balsam alone, and consider that the labour is much lessened thereby.

26. A description of that mode of injection which is most generally employed has now been given, but this is not the only method of effecting our object. A most ingenious process was invented by M. Doyers, requiring no artificial warmth, by which many beautiful objects have been prepared. Make a solution of bichromate of potash, 524 grains to a pint of water, and throw this into the vessels to be injected; then take 1,000 grains of acetate of lead dissolved in half a pint of water, and force this into the same vessels. A decomposition now takes place in the vessels, and the yellow chromate of lead is formed. In this decomposition,

however, the acetate of potash also is formed and as this salt has an injurious action upon the cells, Dr. Goadby recommends nitrate of lead to be used, which preserves rather than destroys them. He also advises the addition of two ounces of gelatine dissolved in eight ounces of water, to eight ounces of the saturated solution of each salt; but with this addition the hot-water bath would be required to keep the injecting fluid liquid.

27. Many of these are best mounted in balsam, in the same manner as those made in the ordinary way; whilst others are best shown when preserved in liquids, for which purpose Goadby's fluid may be employed.

28. This mode of making injections with chromate of lead is deemed by many the best, especially where one colour only is employed. But it must be allowed that there is a little more danger of failure where two separate fluids are used for the same vessels.

29. We will now consider the best manner of making *transparent* injections, which, for many purposes, possess an undoubted advantage over opaque ones. But it must be remembered that there are certain subjects to which no transparent injection could be applied, as they are too thick when in their natural state, and cutting would destroy all that beauty which is shown by the different parts in their relative adaptation. For those objects, however, which must be cut into sections to display their wonders, or are naturally thin—such as some of the finer tissues, livers, kidneys, &c.,—transparency is a great acquisition, and enables us to understand the arrangement of the vessels more perfectly. Again, another advantage is the simplicity of the process; no hot water is needed with some preparations, either for the subject or the injecting fluid, which runs into the minute vessels thoroughly and easily, whilst the cost is small.

30. For this kind of injection no colour is so commonly used as Prussian blue. It is not a good one, as was before stated, for any opaque object, the light reflected from it

appearing almost black; yet by transmitted light no colour is more useful, because its distinctness is equally great by artificial light and ordinary daylight. The method of preparing this, as given by Dr. Beale, is as follows:—

Glycerine	1 ounce.
Wood naphtha, or pyroacetic acid	$1\frac{1}{2}$ drachm.
Spirits of wine	1 ounce.
Ferrocyanide (yellow prussiate) of potassium	12 grains.
Tincture of sesquichloride of iron.........	1 drachm.
Water	4 ounces.

Dissolve the ferrocyanide of potassium in one ounce of the water; add the tincture of sesquichloride of iron to another ounce. Mix these solutions gradually together, shaking the bottle well which contains them—it is best to add the iron to the potash solution. When thoroughly mixed, these solutions should produce a dark-blue mixture, perfectly free from any perceptible masses or flocculi. Next mix the naphtha and spirits of wine, and add the glycerine and the remaining two ounces of water. This must now be slowly mixed with the blue liquid, shaking the whole well in a large bottle whilst the two come together. The tincture of sesquichloride of iron is recommended, because it can always be obtained of a uniform strength.

31. Dr. Turnbull used a mixture slightly different from the above, which is made with the sulphate of iron:—

Purified sulphate of iron	10 grains.
Ferrocyanide of potassium	32 grains.
Glycerine	1 ounce.
Pyroacetic acid	$1\frac{1}{2}$ drachm.
Alcohol	1 ounce.
Water	4 ounces.

Dissolve the sulphate of iron in one ounce of the water,

gradually add the ferrocyanide of potassium dissolved in another ounce, and proceed as before.

32. Dr. Beale also gives us the following carmine injection to be employed in the same way as the blue.* Take—

Carmine....................................	5 grains.
Glycerine, with 8 or 10 drops of hydrochloric acid	$\frac{1}{2}$ ounce.
Glycerine (pure)	1 ounce.
Alcohol	2 drachms.
Water	6 drachms.

Mix the carmine with a few drops of water, and when well incorporated add about five drops of liquor ammoniæ. To this dark-red solution about half an ounce of the glycerine is to be added, and the whole well shaken in a bottle. Next, very gradually pour in the acid glycerine, frequently shaking the bottle during admixture. Test the mixture with blue litmus-paper, and if not of a very decidedly acid reaction, a few more drops of acid may be added to the remainder of the glycerine and mixed as before. Lastly, mix the alcohol and water very gradually, shaking the bottle thoroughly after adding each successive portion till the whole is mixed. This fluid may be kept ready prepared, and injections made very rapidly with it.

33. The method of making injections with these colours is the same as with the gelatine mixtures before described, except that no heat is required, and consequently most of the trouble removed. The bottle of the fluid must be well shaken immediately before use; and when the object has been injected, we must allow it to remain in a cool place for a few hours before cutting it. Thin sections of the subject may be cut with Valentin's knife, as before described, and

* When, however, it is desirable to cut very thin sections of the injected subject, the carmine is sometimes added to a solution of fine gelatine—gelatine one part to water eight parts. But the warm water and mode of proceeding which are used with the size solutions before described will be necessary in this case also.

are very beautiful transparent objects. Some of the finer tissues, also, are shown much better by this mode of injection than by the opaque, and are easily mounted by washing in clean water when first separated, and floating upon a slide, where they must be allowed to dry thoroughly. They may then be immediately mounted in balsam, or kept in the dry state until it is convenient to finish them; but in many cases this keeping, if too much prolonged, will injure the object. If it is desired to transfer the section to another slide, it will be necessary to wet it thoroughly with water by the aid of a camel-hair pencil, and then gently strip it off with the forceps. When it is wished to preserve injected subjects in masses, it must be done by immersion in spirit, and the sections may be cut at leisure. Most of these transparent objects may be mounted in Canada balsam; but some recommend glycerine or glycerine jelly, as allowing the use of a higher power in their examination, and preserving them in a more natural form.

34. A few subjects may be noticed which are very beautiful when injected, and amongst these are the eyes of many animals. They must be injected by the artery in the back part, and when the blue transparent liquid is employed, nothing can exceed the delicate beauty displayed by some of the membrane. It must be dissected with care, but well repays us for the trouble. Water-newts and frogs are not difficult subjects, and in their skin and other parts are many interesting objects. Amongst the commoner animals—rats, rabbits, cats, &c.—almost endless employment may be found, making use either of portions or the whole animal at once. The intestines of many of these are very beautiful. We must divide them with a pair of scissors along the tube, and cleanse them from the contents; the specimen may then be laid upon a slide, and any remaining impurity removed by a camel-hair pencil and water. When dried it may be mounted in balsam, and having been injected with the transparent blue, its minute beauty is shown most perfectly. In injecting a sheep's foot, which is a good object,

the liquid should be forced into it until a slight paring of the hoof shows the colour in the fine channels there. Perhaps one of the most beautiful and interesting objects is the skin or section of a cat's tongue. On examination we shall readily learn the reason why we feel such a roughness when we allow the cat to apply her tongue to our hands. In appearance we shall be almost ready to say that there is no little resemblance in the arrangement betwixt this tongue and those of the molluscs already described. The liver of a rabbit or any other animal is an easy and beautiful object for injection. Sections made with Valentin's knife, and mounted in balsam, are not at all difficult, and worth double the time they occupy.

35. When the lungs of small animals are injected, the finest fluid must be used, as some of the capillaries are so small that it is not an easy matter to fill them properly. And before entering upon these subjects, a certain proficiency in the mode of using the syringe, &c., should be obtained by practising upon simpler parts.

36. No subjects are more difficult to inject than fish, owing to the extreme softness of their tissues. Mr. Hogg recommends the tail of the fish to be cut off, and the pipe to be put into the divided vessel which lies just beneath the spinal column; by which method beautiful injections may be made. The gills, however, are the most interesting part as microscopic objects.

37. These instructions may seem very imperfect to those who have had much experience in this branch; but they will remember that their own knowledge was not gained from any written descriptions, but was forced upon them by frequent failures, some of which probably were very disheartening. As I before stated, it is very difficult (if not impossible) to accomplish much without some knowledge of anatomy.

38. I may here mention that the transparent injections sent over from the Continent are beautifully executed by Hyrtl of Vienna (who states that the injected fluid is com-

posed of gelatine and carmine), Dr. Oschatz of Berlin, the Microscopic Institute of Wabern, Schaffer and Co. of Magdeburg, and others. Some of these will bear examining with a high power. A friend informs me that he measured a vessel in a rat's tongue by Hyrtl, which was 1-7200th of an inch in diameter, and had a clear outline with quarter-inch objective. He has also made many experiments with the same materials, but has as yet failed in producing perfectly distinct outlines, there being a tendency of the colouring matter (magenta, carmine, &c.) to diffuse itself through the coats of the vessels into the surrounding tissues, although he has varied the pressure from half a pound to sixty pounds. He believes the vessels are first washed out (injected with warm water and pressure applied), then some fluid introduced, probably solution of tannin, which renders the arteries impervious to the coloured fluid afterwards injected.

39. He finds that after washing out the vessels as above, the injecting fluid is much more easily introduced. He has used a strong solution of gallic acid previously to injecting with the colouring matter (in one experiment only), and the result was satisfactory. He puts the query,—Might not carbolic acid have a similar effect? He has often used it with injections to preserve the specimens, but not in sufficient quantity to act in the way indicated above.

Since writing the above, Mr. J. G. Dale, F.C.S., and I have made numerous experiments with carmine injection, and have at length been favoured with what we deem success. Some of the vessels in a kitten lately injected do not exceed 1-2000th of an inch in diameter, and present a clear outline with one-fifth objective. There is no extravasation, neither does the colouring matter show any grain except when a very high power is employed. The following is our process :—

Take 180 grains best carmine.
$\frac{1}{2}$ fluid ounce of ammonia, commercial strength, viz., 0·92, or 15° ammonia meter.
3 or 4 ounces distilled water.

Put these into a small flask, and allow them to digest without heat from twenty-four to thirty-six hours, or until the carmine is dissolved. Then take a Winchester quart bottle, and with a diamond mark the spot to which sixteen ounces of water extend. The coloured solution must be filtered into the bottle, and to this pure water should be added until the whole is equal to sixteen ounces.

Dissolve 600 grains potash alum in ten fluid ounces of water, and add to this, under constant boiling, a solution of carbonate of soda until a slight permanent precipitate is produced. Filter and add water up to sixteen ounces. Boil and add the solution to the cold ammoniacal solution of carmine in the Winchester quart, and shake vigorously for a few minutes. A drop of this placed upon white filtering-paper should show no coloured ring. If much colour is in solution the whole must be rejected, because, although it is possible to precipitate all the colouring matter by the addition of ammonia or alum, it is not well to do so, as the physical condition of the precipitate is thereby altered.

Supposing the precipitation to be complete, or very nearly so, shake vigorously for at least half an hour, and allow it to stand until quite cold. The shaking must then be renewed for some time, and the bottle filled up with pure water.

After allowing the precipitate to settle a day, draw off the clear supernatant fluid with a syphon. Repeat the washing until the clear liquid gives little or no precipitate with chloride of barium. So much water must be left with the colour at last that it shall measure forty fluid ounces.

For the injecting fluid, take 24 ounces of the coloured liquid thus prepared, and three ounces of good gelatine. Allow these to remain together twelve hours, and then dissolve by the heat of a water-bath; after which it should be strained through fine muslin.

As this injecting fluid contains gelatine, the hot water, and other contrivances mentioned in a former part of the chapter, will be necessary here also, but no peculiar treatment will be required.

Since writing the above concerning *carmine injection*, I have had the misfortune to lose my friend Mr. Dale, but not before we made scores of experiments together, with this formula. Our experience, I may say without vanity, justifies me in declaring that a good operator can get results equal to any that he will receive from the Continent, as far as colour and distinctness are concerned. The colour, being thoroughly precipitated, cannot *stain* the tissues, and the course is thus clear and well defined. If the object is small, it is well to use the mixture with 25 or 30 ounces of water instead of 40; but, with this exception, I know of nothing that needs alteration. I have, time after time, measured vessels thoroughly filled with good colour — especially amongst fatty portions,—and found them betwixt 1-3000th and 1-4000th of an inch. As some young students might say " Give me an account of something done with it," I will endeavour to describe my use of part of a *horse's leg*. My friend Mr. Hepworth wrote to me that he had a horse's leg, and should be glad if I would come over with sufficient carmine to inject it. I took three pints of solution, and may here mention that, with a very slight loss indeed, all this liquid was thrown into our subject. The leg was cut off just beneath the knee, and before using it we allowed it to remain in water about 80° or 90° Fahrenheit two or three hours, and then introduced our syringe into an artery at the top. As I have no faith in any mechanical contrivance,

I used the common syringe, and filled the leg with the liquid. We then placed it in cold water, and allowed it to remain until the next morning.

The first work was to remove a piece of the skin, and take sections of it by the aid of a Valentin's knife. The arteries were shown beautifully; but the most attractive part was where the growth of the hair was laid open before us. Each hair exactly resembled a common onion, whilst every bulb was surrounded by a perfect network of arteries; and where any bulb had been torn out by accident, there was left a minute *bird's-nest* of them, clearly showing how they had been intertwined around some lost friend. Different sections of the muscular portions showed every phase of arterial distribution, with some exquisitely minute vessels in parts. I then took an artery and cut cross sections, in which the carmine portions were as closely interwoven as wickerwork. I also, with a pair of scissors, laid open a length of the artery, and mounted it, together with a cross section. In the same way I used the veins. Many of the nerves we took out, and, after cleaning carefully with knives and small brushes where necessary, mounted them with the attendant arteries around them. But as we approached the hoof, double interest was given us. The skin with its hairs just above the hoof plainly showed the change taking place, and sections of the hoof gave beautiful specimens of where circulation was gradually stayed by the growth of harder substance. Here, too, we reached the laminæ, or thin plates (somewhat resembling the gills of certain fishes), the exact use of which, we have no space to discuss—and these were readily removed by the aid of scissors and knife. In these the vessels are minutely and exquisitely shown.

These are a few of the beauties which this injection afforded me. My friend Mr. Hepworth and I worked together at this subject for a week or two, and part of the knowledge which he gained from it was communicated to the microscopic world in the fifth volume of their Journal, where the illustrations are beautifully printed. We took about 1,000 slides from this leg, but could easily have taken a specimen for every microscopist in the country. What few slides I now have are mostly mounted in balsam, and are quite as good in colour and every way as beautiful as on the day they were mounted.

CHAPTER VIII.

MISCELLANEOUS.

It must be evident to all readers that there are various objects of interest to the microscopist which cannot be properly placed amongst any of the forementioned classes, but must not be omitted in such a guide as this professes to be. Of these may be mentioned the circulation of the blood in various animals, the rotary motion of the fluid in many plants, the best means of taking minute photographs, &c. &c.

Perhaps the most interesting of these objects is the circulation of the blood through the finer vessels of various parts of animals used for these purposes, which parts, it is evident, must be very transparent to afford a perfect view of this phenomenon. The web of the frog's foot is very frequently employed, but requires a certain arrangement, which we will now describe. A piece of thin wood (Dr. Carpenter recommends *cork*) is taken, about eight inches long and three wide; about an inch from one end is cut a hole, half or three-quarters of an inch in diameter. The body of the frog is then placed in a wet bag, or wrapped in wet calico, whilst the hind foot projects; the whole is then laid upon the piece of wood, so that the foot, which is left free, may be extended over the hole. The web must then be spread out, and secured, either by loops of thread fastened to the toes and attached to small pins placed around the hole in the wood, or the pins may be inserted into the wood—through the web. A few bands of tape must be passed round the body, the leg, and the wood, to prevent any disarrangement arising from the struggles of the animal. Care must be taken that the tape be not too tight, else the circulation will be very slow or altogether stopped. The wood must now be fixed upon the stage, with the aperture under the object-glass: this is sometimes done by simply binding it, or a spring is fixed so as to accomplish the same object without so much trouble. With a half-inch power the blood may be seen to flow very distinctly. The frog may be used for hours if care be taken to prevent the web from becoming dry, by wetting it with a little water from time to time. The piece of wood or cork upon which the frog is laid is often made to give place to the "frog-plates," supplied by opticians. These are made of brass, somewhat resembling the piece of wood above recommended, but each maker's pattern differs, according to his own taste.

The tongue of the frog is also sometimes used for the purpose of showing the circulation of the blood, which is done in the following manner:—The body is wrapped with calico, and made fast to the plate as before, only the *mouth* of the frog is brought to the opening. The tongue is then gently drawn out of the mouth and pinned down over the aperture, when the circulation will be well shown. But, as Dr. Carpenter observes, the cruelty of this mode of treatment is so repulsive that it is unjustifiable.

Tadpoles of the frog (which, of course, are only obtainable in their season) are good subjects for showing the circulation of the blood. They are best suited for the microscope when about one inch long. The tadpoles of the newt and toad also are equally suitable. They may be placed in a very shallow glass trough with a little water, and a narrow band of linen bound lightly round in some part not required for examination, to keep them from moving; or they may be laid upon a glass plate with a drop or two of water, and a thin glass covering lightly bound upon it. Dr. Carpenter, however, places them first in cold water, gradually adding warm until the whole becomes about 100°, when the tadpole becomes rigid, whilst the circulation is still maintained. I have not, however, found this necessary, the thin glass accomplishing all that is desired. The tail is generally the most transparent, and shows the circulation best; but in some of the newt larvæ the blood may be traced down to the very extremities if they are not *too old.* Mr. Whitney places the tadpole upon its back, by which means the heart and other internal arrangements may be seen.

Amongst fishes also may be found subjects for the same purpose, but they seldom furnish such good examples as those before mentioned, because the blood-vessels are not nearly so abundant as in the foot of the frog, &c. The stickleback is, however, procurable almost in any place during the summer months, and may be laid in a shallow trough, loosely bound down as the tadpole. The tail may be covered with a piece of thin glass to prevent him curling it to the object-glass. The power needed for this will be about the same as with the other subjects; viz., a half to a quarter-inch object-glass.

It is not absolutely necessary to go to reptiles or fishes for this curious sight, as some other animals serve very well. In the wings of the common bat may be found a good subject. These must be stretched out on something resembling the frog-plate before described, when those parts near to the bones will show the *largest* vessels very clearly. The ear of a young mouse is an illustration of the same phenomenon, but it is very difficult to fix it in a good position, as these animals are so very timid and restless.

Amongst insects also the circulation may be seen by placing them in the cage, or live-box, so as to keep them still, but not to injure them by too much pressure. In certain larvæ it is particularly well shown, as in those of the day-fly and plumed gnat; but in some of these the blood is almost colourless. In the wings also of many insects this circulation is well seen, as in those of the common housefly; but as these parts become dry in a few days, the subject should not be more than twenty-four hours old.

Somewhat approximating to the forementioned phenomenon, is the rotation (or cyclosis) of fluid in the cells, or, as it is usually termed, the *circulation of the sap*, of plants. This is shown in certain vegetable growths as a constant stream of thick fluid, wherein small globules are seen; which stream flows round the individual cells, or up the leaf, turning at the extremity, and down again by a different but parallel channel. There is little or no difficulty in showing this in many plants; but some are, of course, better than others, and require a different treatment; we will, therefore, notice a few of these. Perhaps the best of all is the *Vallisneria spiralis*, which is an aquatic plant, frequently grown in, but not really belonging to, this country. As it somewhat resembles grass, the leaf is not used in its natural state, but a thin section cut lengthwise with a razor or other sharp instrument; this section, however, is much better when the outer surface has been first removed. It should then be laid upon a slide with a drop or two of water, and covered with a piece of thin glass. Often the cutting of the section seems to be such a shock to the leaf that no motion is visible for awhile, but in a short time the warmth of an ordinary sitting-room will revive it, and with a quarter-inch object-glass the currents will be rendered beautifully distinct. Where the stream is unusually obstinate, the warmth may be slightly increased, but too high a heat destroys the movement altogether. In the summer, any of the leaves show this circulation very well; but in the winter, the slightly yellow ones are said to be the best.

The Vallisneria requires to be cut in sections to show this circulation; but there are many plants of which it is but necessary to take a fragment and lay it upon the slide. The *Anacharis alsinastrum* is one of these: it grows in water, having three leaves round the stem, then a bare portion, again another three leaves, and so on. One of these leaves must be plucked close to the stem, and laid upon a slide with a drop of water. Thin glass should be placed upon it, and along the mid-rib of the leaves the circulation may be seen most beautifully when a good specimen has been chosen; but it requires rather more power than the Vallisneria. This plant is very common in many parts of the country, a great number of our ponds and streams being literally choked up by it. In the *Chara vulgaris* and two or three of the Nitellæ, &c., this phenomenon may also be seen with no preparation except plucking a part from the stem and laying it upon a slide, as with the Anacharis. In using the Frog-bit, the outer part of the young leaf-buds must be taken to obtain the best specimens for this purpose; but a section of the stem will also show the circulation, though not so well. The plants before mentioned are all aquatic, but the same movement of the globules has been observed in several kinds of land plants; as in the hairs upon the leaf-stalks of the common groundsel; but these do not show it so well, nor are they so easily managed as the preceding.

Many microscopists who are not fortunate enough to be in the neighbourhood of these plants (indeed the Vallisneria is a foreign one) grow them in jars; so a few remarks as to the treatment they require will not be out of place. The Vallisneria requires a temperature not lower than 55° or 60°, and even a higher degree than this renders its growth quicker; and no great change must take place: the more equable the temperature the more healthy will the plant be. A glass jar should be taken, having an inch or two of mould at the bottom, which must be pressed down closely, and the plant must be set in this. Water must then be gently poured in, so as not to disturb the mould. As this plant flourishes best when the water is frequently changed, Mr. Quekett recommends that the jar should be occasionally placed under a tap of water, and a very gentle stream allowed to fall into it for several hours, by which means much of the confervoid growth will be removed and the plant invigorated. The Anacharis may be rooted in the earth like the Vallisneria, but a small detached piece may be thrown into the jar of water and there left until wanted. For months the circulation will be well shown by it, and it will probably grow and increase. It is also very healthy in an in-door aquarium. It is recommended that the jars in which the plants of *Chara* are grown should be moved about as little as possible, as the long roots are very tender, and will not bear agitation.

An object which is interesting to the microscopist, as well as the unscientific observer, is the *growth* of seeds, as it is often erroneously termed. A shaving of the outside of the seed is taken and laid upon the glass slide; a thin glass cover is then placed upon it, and a drop of water applied to its edge. The water will gradually flow under the glass and reach the section of the seed, when transparent fibres will appear to spring out and "grow" for some minutes. This, however, is produced by the unfolding of a spiral formation in the cells, and, therefore, has really no similarity to the true growth. The seeds of the Salvias, Collomias, Senecio, Ruellia, &c., are well suited for the display of this curious sight.

To watch the development of the spores of ferns, and the fertilization and products, Dr. Carpenter recommends the following mode of proceeding:—"Let a frond of a fern, whose fructification is mature, be laid upon a piece of fine paper, with its spore-bearing surface downwards; in the course of a day or two this paper will be found to be covered with a very fine brownish dust, which consists of the discharged spores. This must be carefully collected, and should be spread upon the surface of a smoothed fragment of porous sandstone; the stone being placed in a saucer, the bottom of which is covered with water, and a glass tumbler being inverted over it, the requisite supply of moisture is insured, and the spores will germinate luxuriantly. Some of the prothallia soon advance beyond the rest; and at the time when the advanced ones have long ceased to produce antheridia, and bear abundance of archegonia, those which have remained behind in their growth are beginning to be covered with antheridia. If the crop be now kept with little moisture for several weeks and then suddenly watered, a large number of antheridia and archegonia simultaneously open, and in a few hours afterwards the surface of the larger prothallia will be found almost covered with moving antherozoids. Such prothallia as exhibit freshly-opened archegonia are now to be held by one lobe between the forefinger and thumb of the left hand, so that the upper surface of the prothallium lies upon the thumb; and the thinnest possible sections are then to be made with a narrow-bladed knife perpendicularly to the surface of the prothallium. Of these sections, which after much practice may be made no more than 1-15th of a line in thickness, some will probably lay open the canals of the archegonia, and within these, when examined with a power of 200 or 300 diameters, antherozoids may be occcasionally distinguished."

Another interesting object to the young microscopist is afforded by the spores of the Equiseta (or Horsetails, as they are often called). These may be obtained by shaking the higher portion of the stems when the spores are ripe. They will then fall like small dust, and may be placed under the microscope. The spores are then seen to consist of a somewhat heart-shaped mass with bands rather intricately curled around it. As they dry, these bands expand, and are seen to be four lines at right angles, with the ends clubbed, as it may be called. If, whilst watching them, the spores are breathed upon, these bands immediately return to their former state, and are closely curled around the spore; but as they gradually dry, again expand. This experiment may be repeated many times, and is a very interesting one.

The preceding are the principal objects which could not possibly be included in any of the former chapters, but would have left a most interesting branch untouched had it been neglected. There is another subject also which should not be passed by; viz., the production of minute pictures which serve as objects for microscopic examination. I may here mention that as this manual is simply to enable the student to prepare and mount his objects, the photography of *magnified* objects has evidently no place here.

Few slides caused so much astonishment as these minute photographs when first exhibited; small spots were seen to contain large pictures, and a page of printed matter was compressed into the one-hundredth part of a square inch. It would be impossible in this place to give the inquirer any instruction in the manipulation of photography, so it must be assumed that he already knows this.

We will first consider the process performed by artificial light. The collodion employed in photographing generally shows as much structure when magnified as is found in linen of moderate texture; but this is not always the case, as some samples bear much enlargement without any of this appearance. It is evident that a structure so coarse would make it entirely unfit for these minute pictures, as all the small markings would be destroyed, or so interfered with, that no great enlargement would be practicable. To obtain almost structureless collodion is not an easy matter, and a clever practitioner in this branch of photography states that he knows of no method to accomplish this with certainty, but he himself tries different samples until he falls upon a suitable one, which he then lays aside for this object. A beneficial effect is often derived from keeping the collodion awhile, but this is not always the case. The slides should be chosen of an equal thickness, so that when focussed upon one, no re-adjustment may be necessary for the others. The glass should, of course, be free from any roughness, scratches, or other imperfections, and of very good quality and colour.

The microscope must be placed in a horizontal position, and the eye-piece removed, the stage having a small clip upon it to keep the prepared plate in position. The negative must be supported at a distance from that end of the microscope tube from which the eye-piece was withdrawn. This distance will, of course, vary according to the relative sizes of the negative and desired picture. With an inch object-glass, which is of a very convenient focus, it will vary usually betwixt one and four feet. The negative must be lighted by an argand gas-burner or camphine-lamp, and the rays rendered as parallel as possible by a large plano-convex lens placed betwixt the light and the negative. It is not easy to arrange the apparatus so as to get the light *uniform;* but a little practice will soon obviate this difficulty. Ordinary ground-glass is too coarsely grained to focus upon, as the magnifying power used to examine the minute reflection must be considerable. One of the slides must therefore be coated with collodion, submitted to the silver-bath, and after washing with water, allowed to dry. Upon this may be focussed the reflected image, and its minuteness examined with a powerful hand-magnifier, or another microscope placed behind in a horizontal position. When the utmost sharpness of definition is obtained, it is usually necessary to remove the plate a little distance from the object-glass, as those for the microscope are slightly over-corrected, so that the chemical rays which accomplish the photography are beyond the visual ones. The exact distance required to give a picture to show the greatest distinctness cannot be given by rule; but experiments must be made at first, and it will always be the same with the object-glass which we have tested.

The plate may now be prepared as in ordinary photography, and placed upon the stage whilst the light is shaded. When all is ready, the shade is removed and the process allowed to go on, usually for thirty or forty seconds; but no certain rule can be given as to the required time, on account of the variety of collodions, lamps, and powers used. It may be here mentioned, that it is well to contrive some little frame to receive the prepared plate, as the silver-bath solution is liable to get upon the microscope-stage, and so, to say the least, disfigure it. When the exposure has been continued sufficiently long, the picture may be developed by any of the ordinary methods, but some of the best productions have been brought out by the aid of pyrogallic and citric acid solution, with the addition of a little alcohol. The fixing may be effected by a strong solution of hyposulphite of soda, and the picture should then be very well washed with pure water. When dry, the photograph must be mounted with Canada balsam, in the same manner as any ordinary object; but great heat must not be used, or the picture may be injured.

When ordinary daylight is employed for this purpose, a dark slide will be required for the prepared plate, in the same way as for photographing landscapes, &c. These dark slides are generally made by each individual to suit his particular arrangements of negatives; but it may be here recommended that the operator should always focus in the same slide which he is about to use, as so small a difference in distance lies betwixt perfection and failure.

For an ordinary student, perhaps the preceding method is that which is the most readily used, and consequently the most generally available; but almost every one has a different arrangement of microscope, by which he procures these minute pictures. Mr. Shadbolt (one of our most successful photographers) gives the following instructions:— "Having removed the upper stage-plate of a large compound microscope, I replace it with one of wood, supplied with guide-pins of silver wire, in order to admit of its supporting a slip of glass coated with collodion, and excited in the nitrate of silver bath in the usual way. If the ordinary brass stage-plate were left undisturbed, it is obvious that it and the excited slip of glass would be mutually destructive.

"The microscope is now to be placed in a horizontal position, the objective, intended to produce the picture, made to occupy the place usually filled by the achromatic condenser on the *sub-stage* of the microscope, while *another* objective is screwed into the lower end of the body of the instrument, which is used not only to focus with, but also to make the requisite allowance for actinic variation.

"The negative intended to be reduced is then arranged vertically, with its centre in the axis of the microscopic body, at a distance from two to four feet from the lower object-glass, and with a convenient screen of card, wood, or thick paper, to cut off any extraneous light that would otherwise pass beyond the limits of the picture.

"A small camphine-lamp is employed for the purpose of illuminating the negative, having a good bull's-eye lens as a condenser, so arranged with its flat side next the lamp that the refracted rays shall just fill the whole of a double convex lens of about six inches in diameter, the latter being placed in such a position as to refract the rays of light in a parallel direction upon the negative. By this arrangement the *bull's-eye lens* of about two inches and a half in diameter *appears* as the source of the light instead of the small flame of the lamp.

"By using a bat's-wing gas-burner of a good size, a *single* lens, instead of the two, may be so placed as to give the necessary uniformity of illumination."

This arrangement requires the same care in working as that before mentioned, the pictures being produced, developed, and fixed by the same treatment.

It is certain that almost every manipulator makes some small changes in the method of producing these minute pictures; but the rules given, though far from new, are sufficient for all purposes; and I may say with truth, that those which I procured when these wonders were quite new, are fully equal in every respect to the best usually met with at the present time.

With these instructions I shall close my Handbook, as I believe that nearly every branch of the Preparation and Mounting of Microscopic Objects has been treated of. Not that the beginner can expect that he has only to read this to be able to mount everything; but that there are difficulties from which he may be freed by instruction, when otherwise he would have been compelled to learn by failure alone. I may here, however, repeat certain advice before given,—

that, when practicable, it is a good thing to mount each object by two or more different methods, as very frequently one feature is best shown dry, another in liquid, and a third in balsam. Secondly, let no failures discourage you in following up what will assuredly one day become a source of great pleasure, and render your daily constitutional walk, which is often dull in the extreme, very delightful, as it will afford you some new wonder in every hedge-row. And, lastly, let the mounting be studied *thoroughly*, scarcely any part of microscopic science being more worthy of thought than this, since it will so far contribute to the enjoyment or instruction of others, as to preserve for their examination, objects which have already ministered to your own, but which may yet be so perishable as to be speedily lost unless some one of the many processes described in this manual be employed for their preservation.

INDEX.

A.

PAGE
ACARIDA 110
Acid, acetic 5
Acid, carbolic 127
Acid, chlorhyd 5
Acid, hippuric 114
Acid, hydrofluoric 19
Acid, hyperosmic 9
Acid, nitric 5, 61, 62
Acid, nitric, with chlorate of potash 6
Acid, oxalic 6
Acid, phosphoric 5
Acid, sulphuric 5, 11
Acid, tannic 5
Agents, "differentiating"... 5
Agents, hardening 9
Agents, softening 13
Agents, staining 7
Air-bubbles 42
Air-bubbles, to expel 91
Albumin.......................... 16
Alcohol 5, 6, 9, 17
Algæ and desmids, to preserve 127
Algæ, to preserve 125
Ammonia 5
Ammonia, molybdate of...... 7
Ammonia, oxalurate of 112
Ammoniæ, liquor............ 14, 67
Anatomical specimens, Brunetti's mode of preparing 164
Aniline colours 7
Animal tissues, dissection of 163
Anise, oil of 7
Antennæ..................... 107, 137
Antennæ, to bleach............ 107
Asphaltum..................... 45, 49

B.

PAGE
BALSAM and Chloroform...... 44
Balsam, Canada 7
Baryta water 6
Bath for injections 177
Bath, hot-water, cheap 93
Beaker glasses 65
Benzole 17, 19
Benzole—Benzine 75
Berg-mehl 72
Blights 87
Blood-discs 83
Blood-stains, to detect 84
Blue, Prussian 8
Boiling 11
Bone, chips of, stained 149
Bone, sections of 148
Bones, fossil, sections of ... 150
Bones, softened 15
Bottle-washing.................. 76
Brain and spinal cord, sections of 166
Breeding-cage for poduræ, McIntyre's...................... 83

C.

CALCAREOUS matter, solvents of 18
Camphor 20
Camphor-water 13, 123
Canada balsam.................. 43
Canada balsam and benzole 90
Carmine solution 7
Casein 16
Cassia, oil of 7
Castor oil 19, 127

PAGE
Castor oil, to mount crystals in ... 113
Cells, built up ... 129
Cells, brass ... 31
Cells, of card ... 29
Cells, epithelial ... 7
Cells, how to fill with balsam ... 95
Cells for test objects ... 32
Cells, ivory ... 31
Cells, leather ... 30
Cells, paper ... 56
Cells, pill-box ... 31
Cells, Piper's ... 31
Cells, tin, zinc ... 31
Cells to fill ... 132
Cells, wooden ... 29
Cellulose in coal ... 144
Cement cells ... 54, 128
Cement, Dr. Bastian's ... 129
Cement, electrical ... 46
Cement, Marion ... 47
Cements ... 43
Cerebral tissue ... 13
Chalk, organisms in ... 98
Chamber, moist ... 22
Chondrin ... 17
Chromic acid ... 9
Circulation in Anacharis and Vallisneria ... 199
Circulation in foot and tongue of frog, to see ... 196
Circulation in insects ... 198
Circulation in tadpoles ... 197
Clips, spring ... 41
Cloves, oil of ... 7
Coal, cannel ... 144
Coal sections ... 143
Colours for injections ... 178
Condenser, cheap ... 51
Coniferous wood ... 144
Corallines ... 81
Coral sections ... 143
Cornea, nerves of ... 159
Covers, glass, to cut ... 26
Covers, ornamental paper ... 57
Covers, glass, to clean ... 28
Covers, to gauge ... 27, 28
Creosote ... 7
Crystallization ... 111

PAGE
Crystallization, compound ... 115
Crystals, sections of ... 157
Cuticles, siliceous ... 121

D.

Dammar ... 44
Dammar and benzole ... 20
Dammar cement ... 45
Deane's compound ... 125
Decalcifying process ... 141
Diamond beetle ... 88
Diatomaceæ ... 58, 60
Diatomaceæ, cleansing of ... 61
Diatomaceæ, collection of ... 59
Diatoms, cleansing of by Mr. Rylands ... 63
Diatoms in situ, to mount by Rylands' method ... 136
Diatoms, to bleach ... 65
Diatoms, to mount ... 63, 67, 69
Diatoms, to mount in Canada balsam ... 97
Differentiation, chemical ... 5
Differentiation, mechanical 5, 7
Digester, Papin's ... 15
Disc revolver, Beck's ... 53
Dissection ... 160
Drying ... 11
Drying tissues, Mr. Suffolk's process ... 159
Dyticus, foot of ... 109
Dyticus, spiracles of ... 109
Dyticus, wings of ... 109

E.

Earth, Bermuda ... 72
Echini, spines of ... 142
Eggs of insects, to preserve 137
Equiseta, spores of ... 202
Equisetum, cuticle of ... 120
Electricity ... 21
Elutriation ... 66
Entomostraca ... 134

PAGE
Ether 19
Eye-pieces, deep 4
Eyes of insects 106

F.

FARRANTS, Mr., his medium 124
Feathers of birds 110
Feet of insects 108
Ferns, fructification of 201
Ferns, scales of 84
Ferns, spores of 85
Fibrin 16
File or Rubber, corundum 140
Filter, tube 77
Finishing 58
Fish, scales of 88
Fish, tails and fins of 84
Flint, sections of 146
Flustra avicularis 119
Foraminifera 73
Foraminifera, sections of ... 78
Foraminifera, to cleanse... 74, 75
Forceps 37
Forceps, bulldog 176
Forceps, compressing...... 38, 39
Forceps, Goode's 39
Forceps, Page's 38
Forceps, Spencer's 41
Forceps, wooden 38
Freezing tissues 11
Frog, cornea of 8
Fruit-stones, sections of...... 150
Fuchsine 8
Fungi and moulds 87

G.

GELATINE 17, 177
Geranium, petal of 80
Gills of fish 168
Gizzards 173
Glass-engraving 19
Glue 177
Glue, liquid 46
Glue, marine 45
Glycerine... 5, 7, 10, 13, 124, 137

PAGE
Glycerine and gum 124
Glycerine jelly 125
Glycerine, refractive power of 14
Glycerine, use of, by Messrs. Suffolk and Hislop 138
Goadby, Mr. 186
Goadby's fluid 126
Goadby's solution 164
Gold, chloride of 8, 159
Gold, chloride of, and potassium 8
Guano, diatoms in 70
Guano, diatoms in, by Mr. Roberts 71
Guano, spurious 72
Gum 7
Gum-water 47
Gum-water, extra adhesive 47

H.

HÆMATOXYLINE 7
Hairs 89
Hairs, sections of 152
Hardening tissue 9
Heat, use of 14
Horn 178
Horn, sections of 150
Horny matters softened...... 14
Horse-hair, plaited for polariscope 118
Hyperosmic acid 10

I.

INDIGO 7
Infusoria, fossil 72
Injection, blue, transparent 187
Injection, carmine 188
Injection, machine for 182
Injection, process of 180
Injections 175
Injections by double decomposition 179
Injections, Dale's method ... 191
Injections, Doyer's method 185

PAGE

Injections, to mount 184
Injections, transparent 186
Injections, twofold 184
Injections, various 189
Insects, bleaching of 21
Insects, examination of, by polarized light 21
Insects, eyes of............... 10, 89
Insects, internal organs of... 17
Insects, legs and feet of...... 88
Insects, muscles of 21
Insects, scales of 81
Insects, to mount in balsam 104
Insects, tracheæ of 92
Inversion of objects............ 50
Iodine................................ 5

J.

JAPAN black 46

K.

KERATIN 17
Kesteven on brain and spinal cord.............................. 12
Kidney, injected 166
Knives, dissecting 37, 161

L.

LABELLING on glass 19
Labels................................ 49
Laminæ, thin 15
Lamp, Bunsen's 42
Lamp, spirit 42
Larvæ, skins of.................. 84
Lathe for sections, Butterworth's 144
Leaves, cuticle of............... 79
Leaves, hairy...................79, 86
Leaves, parenchyma of 17
Leaves, sections of 154
Lenses, immersion 23
Lieberkühn 89
Light, polarized 20

PAGE

Lime-water 6
Liver, injected 166
Lung, injected 166

M.

MACERATION 13, 15, 162
Mayflies 138
Mercury bichloride............ 9, 11
Mercury nitrate 17
Metal capsules for Canada balsam 94
Microscope, dissecting 160
Microscope, Gairdner's 64
Microscope, inverted 22
Mosses and Algæ 133
Mosses, fructification of...... 85
Mosses, to preserve............ 125
Mounting, dry 52
Mounting in Canada balsam 90
Mouth of insects 109
Muscle 164
Muscular fibre 15

N.

NAILS, softened 15
Needles, curved 176
Needles, dissecting............ 161
Needles, use of.................. 37
Nerve-tissue 13, 165
Neutral media 2
Neutral tints..................... 20
Nitric acid......................... 80
Nummulites 17

O.

OATMEAL, parched 11
Objectives of wide angle ... 4
Objects small, to take up... 36
Oils, essential 19
Oily and fatty matters 19
Oily and fatty matters, solvents of 19
Onion, raphides of 120

PAGE
Osmic acid... 8
Ossein... 15, 17
Oysters, young... 117

P.

PALATES of molluscs ... 169
Paleæ of cereals ... 121
Palladium, chloride of... 10
Palladium, proto-chloride of 8
Paraffine ... 12, 158
Pencils, camel-hair ... 36
Photography, microscopic... 203
Picric acid... 7, 8
Pipette ... 66
Podura, scales of ... 82
Polariscope, to mount objects for ... 111
Pollens ... 80
Polycystinæ ... 99
Polycystinæ, to cleanse ... 100
Polyzoa ... 135
Porcupine, quill of ... 151
Potash, bichromate of ... 9, 10
Potash, chlorate of ... 5
Potassæ, liquor ... 5, 13, 14
Potassium, ferro-cyanide ... 8
Potassium, iodide of ... 8
Preservative liquids... 123
Pump, air ... 42

R.

RAPHIDES ... 86
Raphides for polariscope ... 120
Retina... 13
Rings, glass ... 130
Rings, iron... 131
Rings, vulcanite ... 131
Rings, zinc and tin ... 131
Rods, glass, pointed ... 37
Rods of cochlea of ear... 10
Rush, sections of ... 154

S.

SALICINE ... 113
Santonine ... 113

PAGE
Saw, watch-spring ... 140
Scales of eel ... 117
Scales of fish ... 117
Scales of plants... 121
Scissors, dissecting ... 37, 161
Sea-mats ... 87
Sea soundings ... 75
Section-cutter ... 153
Sections ... 139
Seeds ... 81
Seeds and pollen ... 110
Seeds, growth of ... 201
Seeds, sections of... 158
Sertularidæ ... 119
Shells, pearly, section of ... 142
Shells, sections of ... 140
Silica, colloid ... 18
Silica, or silicic acid ... 18
Silica, vegetable ... 67
Siliceous matter, solvents of 18
Silver, nitrate of ... 8
Size, gold ... 45
Skin ... 17
Skin, sections of ... 156
Slides, concave ... 130
Slides, glass ... 24
Slides, gauge for ... 25
Slides, to clean... 25, 94
Slides, to cover and ornament ... 57
Soda, caustic... 9
Sodæ, liquor... 5, 13, 14
Soda, tartrate of ... 114
Soft tissues, sections of ... 155
Softening tissues ... 13
Spicer, the Rev. W. W.'s preservative fluid... 127
Spicula of sponges ... 103
Spider, poison-glands of... 168
Spinal cord ... 165
Spiracles ... 168
Spleen, injected ... 166
Sponge, sections of ... 156
Stage of microscope, hot ... 22
Staining... 7
Staining, double ... 7
Staining nerves ... 159
Stand, Mr. Loy's ... 51
Stand, universal ... 50

PAGE
Starches ... 119
Star-fish ... 78
Stomach of molluscs ... 170
Sublimate, corrosive ... 17
Sugar ... 7
Syringe ... 175
Syringe for balsam ... 93
Syringe, glass ... 162
Syrup ... 14

T.

TABLE, brass ... 42
Tannic acid, or Tannin ... 6
Tannin ... 9, 11, 17
Teasing by needles ... 3
Teeth, sections of ... 92, 146
Teeth, softened ... 15
Tendon ... 17
Testæ ... 18
Thwaite's liquid ... 126
Tin tubes, collapsible ... 44
Tongues of molluscs ... 169
Tongues or palates of mollusca ... 118
Tracheæ of insects ... 109, 167
Trough, dissecting ... 162
Tubes, dipping ... 37
Turntable, Mr. Hislop's ... 33
Turntable, Dr. Matthews' ... 34
Turntable, Mr. Shadbolt's ... 33

PAGE
Turpentine ... 7, 19
Turpentine, spurious ... 91

V.

VALENTIN'S knife ... 155, 184
Varnish, black ... 48
Varnish, caoutchouc ... 48
Varnish, sealing-wax ... 48
Vegetable tissues, dissection of ... 162

W.

WATCH-GLASSES ... 41
Water-bath ... 22
Wine-glasses, broken ... 42
Whalebone ... 14
Whalebone, sections of ... 151
Wood, fossil ... 144
Wood, sections of ... 152

Z.

ZINC, chloride of ... 126
Zoophytes ... 87
Zoophytes, to mount ... 103
Zoophytes, to preserve, by Dr. Bird's method ... 101

ROBERT HARDWICKE, PRINTER, 192, PICCADILLY.

TO STUDENTS OF NATURAL HISTOR[illegible]D AND YOUNG

A NEW NATURAL HISTORY PERIODICAL.

G. P. PUTNAM'S SONS will hereafter publish for the United States an American edition of the well known English Periodical,

SCIENCE-GOSSIP,

An Illustrated Medium of Interchange and Gossip for Students and Lovers of Nature.

EDITED BY J. E. TAYLOR, F. G. S.,

Author of "Geological Essays," "Geological Stories," etc., etc.

TREATING OF SUCH SUBJECTS AS

ANIMALS, AQUARIA, BEES, BEETLES, BIRDS, BUTTERFLIES FERNS, FISH, FOSSILS, LICHENS, MICROSCOPES, MOSSES, REPTILES, ROCKS, SEAWEEDS, WILDFLOWERS, ETC., ETC.

It will be published monthly at the price of 20 cents a number; $2.25 a year. Of the American issue the first five numbers for 1873 are now ready.

Subscriptions can commence in January or July. Specimen numbers sent on receipt of 15 cents.

The volume for 1872, in a handsome American dress, forming a complete and attractive volume on Natural Science and History, copiously Illustrated, is also ready for delivery. Price, prepaid by mail, $2.50.

The six volumes prior to 1872, can also be obtained.

Price, bound in cloth, each - - - - -	$2.50.
Or six volumes bound in two, in half morocco, -	10.00.

This Magazine has achieved in England a great success, and the publishers are assured that its republication in this country will be welcomed by thousands of readers, students and lovers of nature.

The articles are prepared by able scientists, and while put into popular language, are guaranteed as being exact and thorough. The illustrations are copious, carefully drawn and handsomely printed, while the moderate price places the work within the reach of all.

We give below some few of the English opinions of the magazine.

OPINIONS OF THE PRESS:

"This is a very pleasant journal that costs but a trifle a month, and from which the reader who is no naturalist ought to be able to pick up a good deal of pleasant information. It is conducted and contributed to by expert naturalists, who are cheerful companions, as all good naturalists are; technical enough to make the general reader feel that they are in earnest, and are not insulting him by writing down to his comprehension, but natural enough and direct enough in their records of facts, their questioning and answering each other concerning curiosities of nature. The reader who buys for himself their monthly budget of notes and discussions upon pleasant points in natural history and science, will probably find his curiosity excited and his interest in the world about him taking the form of a little study of some branch of this sort of knowledge that has won his readiest attention. For when the study itself is so delightful and the enthusiasm it excites so genuine and well directed, these enthusiasms are contagious. The fault is not with itself, but with the public, if this little magazine be not in favor with a very large circle of readers."—*Examiner.*

"It is a capital idea. Scarcely any periodical of the day ought to be more popular; and if it answers its design, it will be more than most a power to refine, to enlarge, and to elevate the mind."—*Nonconformist.*

"We augur great success for this much-wanted monthly. It is illustrated with excellent wood engravings."—*Keene's Bath Journal.*

"The variety of information it contains on the most interesting subjects is really astonishing."—*Kilkenny Journal.*

"It abounds with interesting information relative to the varied forms of animal life."—*Lincolnshire Guardian.*

"The publication before us is certainly one of the cheapest and best of its kind. Its distinguishing merit is the easy gossiping style in which it treats of scientific topics."—*Staffordshire Sentinel.*

"Students and lovers of nature will find this little book invaluable to them."—*Taunton Courier.*

MICROSCOPES

AND

Microscopic Accessories.

Full line always in stock, ranging in price from 50 cents to $3,000. Also, a full assortment of

MICROSCOPIC OBJECTS.

Catalogue sent by mail to any address on receipt of 10 cents.

JAMES W. QUEEN & CO.,

601 Broadway, N. Y. ***924 Chestnut St., Phila.***

www.ingramcontent.com/pod-product-compliance
Lightning Source LLC
LaVergne TN
LVHW050529100826
845148LV00002B/491

* 9 7 8 1 4 2 5 5 1 9 3 9 1 *